D0699693

636.2
Cooper
1984

Cooper, Malcolm
Profitable beef production

DATE DUE

Return Material Promptly

PROFITABLE BEEF PRODUCTION

PROFITABLE BEEF PRODUCTION

M. McG. COOPER and
M. B. WILLIS

DISCARD

FARMING PRESS LIMITED
WHARFEDALE ROAD, IPSWICH, SUFFOLK

Northeastern Junior College
LEARNING RESOURCE CENTER
STERLNG, CO 80751 57390

First Published 1972
Second Edition 1977
Third (Revised) Edition 1979
Reprinted 1981
Fourth Edition 1984

ISBN 0 85236 154 8

© FARMING PRESS LTD, 1972, 1984

All rights reserved. No part of this publication may be reproduced, stored in a retrieval system, or transmitted, in any form or by any means electronic, mechanical, photocopying, recording or otherwise, without the prior permission of Farming Press Limited.

Printed and bound in Great Britain by
Spottiswoode Ballantyne Ltd, Colchester

CONTENTS

ILLUSTRATIONS

8

FOREWORD

by DR DAVID ALLEN,
Head of Beef Improvement Services,
Meat and Livestock Commission

A CURIOUS coincidence brings together two authors, the senior of whom recently left Britain to take up an important research and development post in Spain, whilst the other recently returned after working on cattle research and development in Cuba. Both have a reputation for controversial comment, yet their book is a restrained attempt to analyse the structure of beef production in Britain and to examine how technical advances can contribute to more efficient and profitable production. Traditional beliefs are not bludgeoned, but persuasive reasoning is used to suggest new approaches to breeding and production. Thus breed improvement is discussed in relation to breed function, breed choice in relation to production systems, and production systems in relation to the individual farm, including the important considerations of capital and cash flow. Some parts of the past are criticised but, with the benefit of hindsight, an important industry should be capable of doing so and learning important lessons for the future.

The authors emphasise economics and production systems and direct their book mainly at those who produce beef. Certainly, on the point of joining the EEC, the book comes at an important time for them. But those in teaching, research and advisory work will also read it in the hope of gaining a better appreciation of the industry they serve.

July, 1972   DAVID ALLEN

PREFACE TO
FOURTH EDITION

THE FIRST edition of this book was published in 1972 just after the United Kingdom had joined the EEC. At that time domestic beef production was at a record level with an annual output of nearly 1·2 million tonnes and generally the beef sector was one of the more buoyant areas of agricultural production. The demand for beef was strong in a favourable economic climate with full employment and rising standards of living and there were high hopes that British beef producers, with a lower cost structure than their Continental counterparts, would profit substantially from access to a greatly expanded market. Unfortunately these hopes have not been realised. Not only has there been a considerable reduction in output, particularly in respect of suckler beef, but appreciable quantities of quality beef are now being taken off the market to go into intervention stores, presumably because of resistance to the target prices that have been set for beef.

What are the reasons for this decline and what are the prospects of it being reversed? There are no easy answers to these questions but unquestionably one factor contributing to the fall in demand was the substantial rise in crude oil prices that took place in the early seventies as the principal oil producers flexed their economic muscles at the expense of consumer countries, which at that time included all the members of EEC. During the years of increasing prosperity the private ownership of cars had greatly increased and they had come to be regarded by a substantial proportion of national populations as an essential adjunct of their prevailing mode of life. There was, in addition, a smaller proportion that also regarded oil-fired central heating as a conventional necessity and to preserve this and the mobility that resulted from car ownership there was a preparedness to forego some of the other good things of life and these included beef.

This situation in Britain was compounded by a substantial rise in beef prices when the country, as a consequence of joining the Common Market, became a dear food country after many decades of relatively cheap food. There were, of course, offsetting rises in salaries and wages that were part of the inflationary situation that Britain experienced throughout the seventies. Thankfully inflation is now at a lower level than it was ten years ago but it is still sufficiently high to cause prudent housewives to opt for poultry rather than beef for Sunday roasts except on special occasions. On top of this there is no longer full employment in any Community country and housekeeping on the dole, the lot of hundreds of thousands of families, is limited to the occasional purchase of mince, never quality beef.

One cannot see any immediate prospect of a reversal of this situation. It is true that the Common Agricultural Policy is examined rather more critically now that there is a greater recognition of its absurdities and the need to do something about them, but it will take a very long time to adjust the imbalances between production and consumption that now characterise the major agricultural commodities. Fortunately there is not such an acute imbalance with beef as there is for many other commodities, notably milk and grain, and economic recovery could quickly obviate the wasteful intervention buying that is unquestionably attributable in the case of beef to under-consumption rather than over-production.

It would be misleading, however, to suggest that a recovery in the demand for beef within the Community would result in great prosperity for producers. Efficiency of beef production in terms of land use, even where it uses surplus animals from the dairy industry, is such that it is not a feasible main line proposition on expensive land which is now the prevailing situation in Britain. The most that one could expect, on all but marginal farms, is for beef production to be an adjunct to other more profitable systems of land use. There is, however, considerable scope for improving efficiency within the various branches of the industry and a principal aim of this book has been to analyse and discuss the factors that determine success in beef enterprises.

In this we have been greatly assisted by the valuable results that have been obtained by the Meat and Livestock Commission and now made available in such publications as their Beef Yearbook. As a result of the Commission's work, covering the whole gamut from enterprise studies, breed assessment and performance testing through to market considerations, there are now facts rather than opinions to provide a basis for planning. We wish to acknowledge the considerable help we have gained from the Commission's publications that have made possible this substantial revision of the earlier edition.

M. McG. Cooper
Holme Cottage, Longhoughton,
Alnwick, Northumberland.

M. B. Willis
Faculty of Agriculture,
University of
Newcastle-upon-Tyne.

July, 1984

ACKNOWLEDGEMENTS

WE ARE grateful to ICI Agricultural Division, Spillers Ltd., *The Irish Farmers' Journal, Dairy Farmer*, and the Milk Marketing Board of England and Wales for permission to reproduce certain photographs—Nos. 3, 4, 11 (ICI), 12 (Spillers), 5 (*Irish Farmers' Journal*), 7, 8, 9, 10 (*Dairy Farmer*), and 13, 14, 15 (MMB). We also wish to thank sincerely Miss Joan Howe who undertook the typing of the original manuscript and to Mrs H. M. Cooper for handling the revisions.

July, 1984
M. McG. C.
M. B. W.

Chapter 1

INTRODUCTION

AFTER THE end of the 1939–45 war it seemed that the traditional roast beef of old England was about to become a memory of more spacious days for the meat ration. Even five years after the end of hostilities consumption was no more than it had been at the height of the war. Supplies from Argentina, the most important pre-war source of imported beef, had virtually disappeared and home production, because of the necessary emphasis on milk and food crops, particularly those for direct human consumption, was at a very low level, both quantitively and qualitatively.

Fortunately for consumers the situation changed dramatically over the following years and by 1970 total supplies of beef were within a few thousand tons of the pre-war level. Per capita consumption then averaged 22 kg which was only 2 kg below the high levels recorded in the late thirties, when slightly more than half of available supplies was home produced. Argentina was then the principal source of imported beef with Australia and New Zealand making significant contributions in the form of frozen beef. Argentine beef on the other hand was mainly chilled and it had a deservedly high reputation for its quality.

Though total supplies, which averaged just on 1·2 million tonnes over the five year period 1978–82, are close to pre-war levels there have been spectacular changes in the sources of supply. Over this period home-produced beef has been responsible for about four-fifths of available supplies and in addition exports of home-produced beef, mainly cow beef, in demand on the Continent for hamburgers, have averaged 120,000 tonnes annually. Argentina for all practical purposes has become a back number and even prior to the Falkland hostilities exports to the United Kingdom were of the order of only 12,000–15,000 tonnes annually. Eire, the principal source of exported beef, sent the United Kingdom 159,000 tonnes in 1980 but since then quantities have fallen because Eire has been developing more lucrative markets outside EEC.

EXPANSION OF HOME PRODUCTION

In recent years domestic production of beef has averaged just over a million tonnes annually and this represents an increase of the order of 70 per cent over immediate pre-war levels of output. There are two principal reasons for this expansion; first, the level of support that successive governments have given the industry and the second, paradoxically, arises from developments within the dairy industry. Not only are there many more dairy cows in Britain than there were forty years ago but also there is a longer herd life which means that a higher proportion of the national dairy herd can be mated to bulls of beef breeds to enhance the supply of calves suitable for rearing for beef. The swing to the Friesian has also been an important factor which will be considered later.

Beef has long been the nation's favoured meat and when supplies were at their nadir expansion of home production became a political issue. Lord Woolton, who had headed the Ministry of Food during the war, recognised this in managing the electoral campaign which returned the Tories to Government in 1951. The slogan then was 'more red meat' and the Government honoured its pledges not only by making beef production a more attractive proposition to farmers but also by taking measures to encourage the rearing of dairy-bred calves for beef. There was no commodity subject to price guarantees that had such generous treatment as beef received at annual price reviews over the following two decades. Between 1951 and 1971 milk and wheat attracted 4·8 and 15 per cent increases respectively whereas the corresponding increase for beef was 114 per cent. Beef also had the benefit of liberal production subsidies, first the headage payment at nine months that was paid on all animals deemed to be potential beef stores and, second, the breeding cow subsidies that were payable on animals kept solely for breeding beef stores.

In the first instance the cow subsidies were limited to stock kept on hill farms but, partly because this involved arbitrary decisions as to what constituted a hill farm, it was decided to introduce a subsidy for beef breeding cows that were not eligible for the hill cow subsidy. The combined effect of the two breeding cow subsidies and the calf rearing subsidy was substantial. In the 1969–70 production year they totalled £44 million or 14 per cent of the total value of output which at that time also included deficiency payments made in respect of price guarantees. Altogether beef made very substantial demands on the Exchequer but looking at this policy from the taxpayers' viewpoint beef was freely available at world prices, which is not the present situation under EEC arrangements.

In 1982 there were over 4·6 million breeding cows in the United Kingdom and of these nearly 1·4 million or 30 per cent consisted of cows kept purely for beef production. This is a very high proportion for a country where there are such pressures on limited land resources, for specialised beef production is an inefficient form of land utilisation. It is essentially a large scale operation of the wide open spaces, for instance the estancias of South America or the ranches of the western rangelands of North America. In most western European countries, particularly Holland, Belgium, Germany and Denmark, beef is primarily a by-product of dairying with a high proportion of cow beef at one extreme of the production range and veal at the other. Unquestionably the United Kingdom would not have its present tally of beef breeding cows if it had not been for the liberal support that the Government gave to the industry from 1951 until 1972 when the country became part of the EEC.

Support for beef production has continued with membership of the Community, for it has been an intrinsic part of the Common Agricultural Policy (CAP) to encourage beef at the expense of milk and milk products which are in a situation of over-supply. Whether this arises from pricing policies or because the people of the Community have all the butter and cheese they want to eat is beside the point; the point being that butter and dried milk mountains have become an expensive embarrassment necessitating a deliberate policy of discouraging milk production. The old type of rearing and breeding subsidies that the British Government made before entry into the EEC have disappeared, but under CAP there are other incentives, some paid by the Community and others by the British Government. There is, for instance, the variable premium paid on higher grade slaughter animals and the two types of cow subsidy. A suckler cow premium was introduced in the 1980–81 marketing year to encourage producers specialising in beef breeding to maintain their herds which were declining because of economic pressure. Payment is limited to holdings where no milk is sold and in 1981 it was £12·37 per cow. It was then for approximately 1·15 million cows or about four-fifths of beef breeding cows.

The second cow subsidy is more substantial and it applies to both beef and dairy cows. It is known as the Hill Livestock Compensatory Allowance and to qualify for it a farmer must operate in what is termed by the EEC to be a 'less favoured area'. It is a subsidy with a considerable social element in its operation, because it is given as a counter against a drift away from the land in the more difficult farming environments of EEC. These incidentally often have a considerable tourist attraction that would be diminished with depopulation. In 1981 the rate of the allowance was £44·50 per cow and it is estimated that 20 per cent of all breeding cows qualified for this payment. The estimate of beef only cows qualifying for this subsidy in 1981 is 64 per cent for the UK as a whole and just over 90 per cent for Scotland, where there is a large area of land classified as 'less favourable'.

IMPORTANCE OF DAIRY-BRED BEEF

Just about the time that Lord Woolton was promising the country 'more red meat', senior members of the Ministry of Agriculture were regretting the eclipse of the Dairy Shorthorn by the Friesian in the national dairy herd on the grounds that the former was a dual-purpose breed while the latter was a dairy breed. Even the secretary of the Friesian Society at that time, perhaps because he was so preoccupied with the clear superiority of his breed in respect of milk yields, insisted that the Friesian was primarily a dairy breed. If at that time the foundation of the breed had been based on the North American Holstein-Friesian rather than the Dutch Friesian there would perhaps have been grounds for this attitude. Fortunately for the future of the beef industry, there were sufficient farmers in the early fifties achieving good results with Friesian stores to convince the National Agricultural Advisory Service that it was high time that it instituted trials to determine the comparative worth of the several kinds of dairy-bred stores that were then available. Those at Rosemaund in Herefordshire were particularly valuable and they clearly established, much to the chagrin of the dwindling band of Dairy Shorthorn breeders, that Friesian steers compared more than favourably with the Dairy Shorthorn, either pure or in the crosses with such beef breeds as the Hereford.

These results were established during the mid-fifties when the great majority of bull calves produced in dairy herds, nearly a million and a half annually, were being slaughtered for 'slink' veal which is mainly used for meat pastes. At that time calf rearing practice was largely based on bucket feeding up to 12 weeks of age and dairy farmers, particularly in the autumn and winter when farm gate prices for milk were at a peak, begrudged feeding the 120–160 litres of whole milk that traditional systems of rearing required. It was far more convenient and rewarding then to sell surplus bull calves as 'bobbies'. Fortunately T. R. Preston revolutionised the hand rearing of calves by studies he undertook at the

Rowett Research Institute over the period 1955–58. Earlier, as a postgraduate at Newcastle, he had established that young calves with access to pasture developed rumen functions as early as one month of age. The logical step, it seemed to him, particularly for calves born in the winter when pasture is not available, was one of offering them a highly digestible palatable concentrate ration from the week-old stage and feeding them limited quantities of milk substitutes up to about 4 weeks of age before switching to an all-solids diet.

The first trial based on these premises exceeded Preston's expectations and his results were soon confirmed by independent work at Cockle Park, which also showed that it was possible to rear calves, that compared favourably with traditional bucket-reared calves, on no more than 100–110 litres of milk and milk substitute plus a high energy supplement with hay as a source of long roughage. The spectacular success of the trials immediately aroused the interest of feeding-stuff firms. Soon leading companies were marketing their versions of the right meal supplement and in the process giving valuable publicity to a system of calf rearing that was appreciably cheaper than conventional systems. More than this it was a system that could be adopted without dependence on a dairy herd or nurse cows, always providing that the calves started life with colostrum feeding to give protection against infections. For those who were prepared to maintain the correct environment for their calves the rearing of good Friesian calves, which could then be purchased as week-olds for as little as £5–6 a head, became a profitable business. Also there was a steadily growing realisation (except in traditional beef areas such as the Borders and further north where the preference was still for Aberdeen Angus crosses) that Friesian steers were potentially more profitable than those of more conventional beef breeding.

DEVELOPMENT OF BARLEY BEEF

Preston was able to give a second boost to the Friesian as a source of beef by his development of a system of production that came to be known as barley beef. Here early-weaned calves are kept on a mainly cereal diet with the necessary supplements, until slaughter at about 12 months when animals combine a reasonable expression of their growth potential with adequate finish. Here the Friesian steer really came into its own as a source of beef, not only because of its high growth rate but also because it is late maturing. This latter aspect is particularly important with high energy diets because early-maturing breeds and crosses quickly run to fat with such feeding and reach slaughter condition at unacceptably low weights, whereas the pure Friesian and its crosses with such late-maturing breeds as the Charolais or Simmental are well suited to the system.

Barley beef production was conceived at a time, early in the 1960s, when the farming industry was very receptive of promising innovations and barley beef was particularly attractive. Good Friesian calves were costing no more than £8 at the week-old stage, or about £25 at 3 months, when they are normally past the early-rearing stage with its attendant problems, while barley was being bought and sold off the combine for £20–22 per tonne. Barley beef became so profitable that it suffered from its own success, since so many were attracted to it that the price of both calves and barley rose to a point where only the really efficient could make a worthwhile profit. This, 20 years later, is still the situation.

Even before the barley beef boom started an alternative strategy in dairy beef production had been developed, namely the 18-month system, which combines intensive utilisation of pasture with limited cereal feeding. Though this results in a longer production cycle and a lowering of efficiency of food utilisation it is nevertheless a logical production system for the United Kingdom which, particularly in its western regions, is so well suited to grass production. Moreover, in a situation where a shortage of potential beef stores is reflected in high prices for rearing calves, it is important to spread this overhead by obtaining a fuller expression of growth potential.

The combined effect of subsidies and incentives on the one hand and the recognition of the fleshing qualities of Friesian and Friesian-cross calves surplus to dairy replacement needs on the other, was probably the major factor in cutting, by over a million, the number of calves that were slaughtered for slink veal. In 1981 there were only 120,000 calves slaughtered at this stage in the United Kingdom. Since the late sixties there have been approximately 1½ million Friesian and Friesian-cross male calves that have been reared annually, that would have been destined for immediate slaughter just after the war. In addition, there have been many Friesian-cross heifers, particularly Hereford crosses, that have been drafted into single suckling herds. It is estimated that approximately 40 per cent of the 'clean' beef produced in the United Kingdom comes from cattle with Friesian dams.

ARTIFICIAL INSEMINATION

Artificial insemination has been an important factor in improving both the quantity and quality of dairy-bred beef. Though there has been a sustained expansion of the dairy cow population in Britain until 1984 there has also been increasing scope since the war for mating a proportion of the national dairy herd to beef bulls and this has been facilitated by the availability of semen from bulls of beef breeds. Approximately one quarter of all inseminations derive from beef bulls and their quality has been progressively improved, especially as a result of selections based on performance tests. The English Milk Marketing Board in particular has been especially concerned with the quality of beef bulls standing at its AI centres. It has also played a notable part in the introduction and testing of Continental beef breeds.

The most popular of the British beef breeds, based on numbers of inseminations, are the Aberdeen Angus and the Hereford. Dystocia is frequently encountered with first-calf Friesian heifers that have been on a high plane of nutrition in late pregnancy and from bitter experience farmers have learned that the Aberdeen Angus is the safest bull breed for first pregnancies, followed by the Hereford. The popularity of the Hereford depends on another factor, namely the strength of demand for white-faced calves, especially those out of the Friesian. The price premium earned by these calves over straight black and whites appears to be growing with the increasing Holstein component of the country's dairy herds, because a Hereford parent will reduce the danger of downgrading of carcass.

When the Charolais was first introduced it was considered that it would be particularly valuable in improving the fleshing quality of dairy-bred stores, but the use of its semen has not become popular with dairy farmers because of calving difficulties and calf mortalities. Even in single suckling herds, where typically cows are on a moderate plane of nutrition prior to calving, there are more than twice as many assisted calvings with Charolais bulls as there are with Herefords, and calf mortality is three times as great with the Charolais. The standard practice of steaming up dairy cows before calving will of course aggravate such calving problems and any gain derived from the increased value of Charolais calves is more than offset by loss of calves and lowered milk yields which are the inevitable consequence of difficult calvings. Unquestionably the place of the Charolais and other large Continental breeds is in single suckling rather than dairy herds.

IRISH CATTLE

Imports of cattle from Ireland have been important in two respects. In the north of England and Scotland the so-called Irish cow is highly regarded by suckled calf producers. Originally this was a cross of the Dairy Shorthorn cow with a beef bull, usually the Aberdeen Angus, but the Shorthorn has been largely superseded by the Friesian and Shorthorn-cross heifers are not easily obtained. So producers now have to accept Friesian crosses sired either by the Hereford or the Aberdeen Angus if they want Irish-bred heifers. Their popularity stems from these attributes: they are hardy, they have frame, they are tractable because they have been hand-reared, and they have sufficient milk to rear a good calf. Home-bred heifers of this kind are now freely available so there is no longer any point in a reliance on imported Irish heifers.

Traditionally, the most important contribution that Ireland made to the United Kingdom's beef supplies was in the supply of well-grown store cattle for short term finishing, either on grass or in yards during the winter months. As recently as 1965 Ireland undertook in a trade agreement to supply 638,000 head of cattle annually. At that time it was estimated that Irish cattle were responsible for about one-sixth of the total home production. Since then, and particularly after both countries joined EEC, the supply of beef stores from Ireland has become progressively smaller and in 1981 and again in 1982 it was estimated by the MLC that only 4 per cent of home-produced beef were derived from stores imported from Ireland. This is a logical development because Ireland has the resources, particularly of grassland, to produce carcasses with a high content of lean meat that meet a strong demand both in Britain and on the Continent. Between 1977 and 1982 Irish beef exported to the United Kingdom has averaged 130,000 tonnes annually and these have accounted for approximately one-eighth of the total supply of beef over this period.

SOURCES OF HOME-PRODUCED BEEF

According to figures published by the MLC, approximately 70 per cent of home-produced beef now comes from home-bred steers and heifers. Of these two-fifths originate in beef breeding herds and the remaining 60 per cent are dairy-bred but with an appreciable proportion of these, at least a quarter, being the progeny of beef bulls. Cull cows and bulls account for approximately one-quarter of the home-produced beef, with dairy animals approximately three times as important as beef animals in this respect. The width of this ratio is a reflection of the appreciably shorter working life of the dairy cow as compared with her beef breeding counterpart.

PRODUCTION AND CONSUMPTION IN EEC

Over the four-year period from 1980 to 1983 inclusive, beef production in the Community has averaged a little under 7 million tonnes annually. The largest producer is France which produces just under 2 million tonnes followed by Germany (1·5 million tonnes) and United Kingdom (1 million tonnes). Total production has had a downward tendency in recent years and in 1982 at 6·65 million tonnes it was 7 per cent lower than it was in 1980. There is a small export trade but this is virtually offset by imports to give a situation where the Community is for all practical purposes self-sufficient.

Because the better grades of beef, as well as veal, are in a sense luxury foods demand for them has been affected by the economic recession that has affected every country in the Community. The tendency has been for pork and poultry to replace the more expensive beef and veal and this has been particularly marked in Germany and the United Kingdom. There is no reason to believe, however, that some recovery in demand would not take place if there is an improvement in the general economic situation.

Looking to the long term future of beef there does not appear to be any prospect of spectacular improvements in efficiency of production, such as those some 30 years ago that transformed poultry from a luxury to being the cheapest. The very nature of beef production is such that there is little scope, certainly in Europe, for economies of scale or for securing improvements of any great magnitude in production techniques to make beef a competitively priced commodity that gives its producers a reasonable living. Indeed with deliberate cut-backs in dairy cattle within the community there could be a reduction in the availability of relatively cheap calves as compared with those that are specifically bred for beef. That is the dilemma that is now facing beef production in the Community.

Chapter 2

SOURCES OF BEEF CATTLE

THOUGH THE European Community is now more or less self sufficient for beef there is little doubt that the existing level of domestic production would be insufficient to satisfy demand if there was a substantial economic recovery. One of the obvious means of making good any shortfall in supplies would be an increase in imports, which over the period 1980–83 averaged just under 400,000 tonnes annually or no more than 6 per cent of total supplies. However there would have to be some radical re-thinking of the Common Agricultural Policy for this to happen. For this reason it is appropriate to consider possible trends in domestic production with particular reference to the situation in the United Kingdom.

BEEF FROM THE DAIRY INDUSTRY

Through a combination of technical advances and Government encouragement the United Kingdom has reached a situation where there can be only a limited scope for any expansion of beef production from the dairy sector. Instead of the one time whole-sale calf slaughtering within a few days of birth, the annual figure is now about 100,000 calves and most of these will either be unthrifty or Channel Island calves with a very limited fleshing capability. There has, however, been quite an appreciable export of calves to the Continent recently, mainly it is believed for rearing as vealers. This trade totalled 420,000 calves in 1978 but in 1982 the figure had fallen to 230,000. Even at this lower level these calves, home-reared for slaughter as 220 kg carcasses, would add half a million tonnes to the country's output of beef. It would also mean an end to a trade that the animal welfare lobby greatly deprecates, sometimes with considerable justification.

The immediate prospect, however, is that the British dairy industry might be providing less and not more calves suitable for rearing for beef, because of EEC decisions to curtail milk production as part of a policy of dealing with embarrassing surpluses of milk products. Any such curtailment could take one of two forms, either a reduction in dairy cow numbers or a reduction in average

24

yield per cow, brought about by a substitution of cheaper forages for more expensive forms of feeding. Some dairy farmers will make one decision and others will choose the alternative course of action, but the inevitable outcome will be fewer calves available for rearing for beef. Two ways of offsetting any fall in numbers would be a reduction in rearing losses, which fortunately are now at a much lower level than they were a few years ago, due to improved hygiene and nutrition, and a reduction in the annual rate of dairy cow wastage, presently about 25 per cent. With a five-year average herd life, which is quite a feasible objective, only 20 per cent of the herd is replaced each year. This means that after making some allowance for wastage, either at birth or later during the rearing stage, nearly 70 per cent of all dairy calves could be available for beef purposes. The present figure, it is estimated, is a little more than 60 per cent. However, reduced herd wastage would mean less cow beef so total beef numbers would be little affected.

There appears to be little scope for increasing the beef potential of calves coming from the dairy industry unless it is a reversal of the recent trend in favour of the Holstein. It is established that it has an excellent growth performance, at least as good as that of the Friesian, but its carcasses are not looked on with much favour by the meat trade. The breeds that make a substantial increase in growth potential when used on dairy cows, mainly the Charolais, Simmental and South Devon, also have the worst records in respect of difficult calvings and neo-natal mortalities. As was pointed out in the previous chapter, a dairy farmer practising steaming up of his dairy cows could be at a financial disadvantage from using big breed semen as opposed to that of a smaller breed, such as the Hereford whose crosses are greatly in demand as rearing calves.

BEEF FROM THE UPLANDS

In the years of acute beef shortage immediately after the war, it almost became a national slogan that we could get more beef from the hills. There were pioneers in this sector, notably Captain Bennet-Evans with Welsh Blacks in Plynlimmon in Mid-Wales and Duncan Stewart with hardy cross-Highland cows on Ben Challum in Perthshire, but in both instances the cattle were associated with substantial sheep enterprises. Bennet-Evans, in particular, was categoric that sheep are masters on the hill. He did not then anticipate that the State would be so liberal in its measures to stimulate beef production and so niggardly in its support for hill sheep up to the point of entry into EEC when a suckler cow

Northeastern Junior College
LEARNING RESOURCE CENTER
STERLNG, CO 80751 57390

rearing her calf on farms like Plynlimmon was earning, in terms of subsidies, as much as 20 Welsh or Swaledale ewes on similar farms. Parity in terms of livestock units should have been one cow to 8 ewes if hill sheep were to have a fair crack of the whip.

These direct subsidies, encouraging an expansion of weaner production, were not the only incentive, for the guaranteed price for beef doubled while sheep-meat prices were virtually stable. It is a little wonder that there was such a substantial increase in the number of suckler cows, not only on the hills but also on adjacent marginal land. Equally it is not surprising that the greater degree of parity between hill sheep and cattle under EEC arrangements and the boost in sheep-meat prices, once some understanding was achieved with France over lamb exports to that country, are resulting in a greater emphasis on sheep production at the expense of breeding cows on upland farms. If the suckler cow premium had not been introduced in the 1980–81 production year there would now be many less suckler cows on Britain's upland farms.

The suckler boom, it must be stressed, was not entirely at the expense of sheep on the majority of hill and upland farms. Increased income from weaners not only demanded better pastures and improved forage conservation but also it provided much of the capital to effect these improvements. There was a spin-off so far as the sheep were concerned, because improved pastures also meant better nutrition for the ewe flock to increase both the size and the quality of the lamb crop.

Leaving aside the differential effects of subsidies and incentives, the debate should be less about the merits of sheep versus cattle on the hill but more on the value of cattle with sheep and the optimal ratios between the two species. On the one hand we have to accept that ewes of the hardy hill breeds have an adaptation advantage over cows, even those of breeds like the Galloway, under the harsh environmental conditions of the uplands. Ewes are under physiological stress for only 6 months of the year as a consequence of their reproductive functions, covering the 2 last critical months of pregnancy and 4 months of lactation. The breeding cow, because of her longer lactation as well as longer gestation, is under stress for at least 8 and possibly 9 months of the year. If one plots the respective nutritive needs of the breeding cow and breeding ewe over the course of a year against the pattern of pasture availability under upland conditions, one sees very clearly where the advantage lies. It does not stop at this, for sheep, with their smaller individual size, their greater agility and their capacity for highly selective grazing, are able to subsist on vegetation that would spell starvation for cows. As a result cattle need more supplementary feeding than sheep in the difficult months of the year and this matters greatly on farms where there is only a limited area of grassland available for conservation.

There are other advantages in favour of sheep but, sensibly managed, cattle have an important place on land where contour and surface conditions prevent the use of mechanical equipment for grassland renovation. The use of cattle as 'animated mowing machines' has long been part of the strategy of grassland management on steep hill farms in New Zealand where sheep are the principal source of income. Indeed by British standards the management of the cattle would seem to be unnecessarily severe, because they have to work so hard for their living for the ultimate benefit of the pasture that their productivity is impaired. It is, however, a question of relative values and in Britain hill farmers could not possibly afford, as their New Zealand counterparts can, to keep bullocks in a store condition until 3 or 4 years of age because they are then more useful than younger animals in their role as tools of pasture management.

TIMING OF CALVING ON HILL FARMS

There is an interesting British compromise between the competing needs for good nutrition for the breeding cows on the one hand and the benefits to be gained from hard grazing of hill pastures on the other hand, which is currently being demonstrated on the Ministry of Agriculture's Experimental Husbandry Farm, Pwllpeiron in mid-Wales. It is on typical Welsh hill land with a short growing season and Nardus as a dominant species in the sward if it is not kept firmly in check. The suckler herd is deliberately calved in July–August with weaning in the early spring. This means that dry cows are available to control Nardus at a very critical time, namely the early spring when it has a measure of palatability, without any detriment to their subsequent performance, for as parturition approaches the cows move to improved pastures and subsequently they and their calves are wintered on silage. It is essential that a farm following such a policy should have sufficient improved grassland not only as a source of silage but also to provide good grazing for weaner calves.

The logical time for calving on most upland farms is in the early spring, about a month before the onset of pasture growth. This is particularly the case where there is limited land available for conservation necessitating an emphasis on quantity rather than quality in the conserved crop. Such timing of calving gives the best coincidence of grassland production with the varying nutritional needs of the cows. Their dry period coincides with the winter and moderate quality hay or silage will suffice until the approach of calving when some improvement in supplementary feeding will be necessary until pasture comes away. Its peak of growth coincides with the period of greatest nutritive demand from the cows and incidentally when it is critically important to have cows on a rising plane of nutrition to safeguard conception rates.

There is one major drawback to April calving and October weaning, namely that seven-month weaners are at a disadvantage as compared with the 9–10 month weaners, of similar breeding, at the autumn sales. The latter can usually be drafted fat out of their winter quarters but the later born weaners are unlikely to reach slaughter condition for another 6 months at least and this is reflected in their selling price. Many upland farmers in consequence have moved to an autumn-calving routine. A comparison of financial and physical performances with the two periods of calving is provided by Table 2.1 summarising MLC survey data for upland herds in the 1982 production year.

TABLE 2.1. COMPARISON OF SPRING- AND AUTUMN-CALVING HERDS ON UPLAND FARMS

	31 Spring-calving herds		33 Autumn-calving herds	
	Average	Top third	Average	Top third
Calf sales per cow (£)	253	273	347	381
Variable costs per cow (£)	84	80	138	138
Gross margin per cow (£)	193	219	232	269
Gross margin per ha (£)	284	422	281	360
Calves reared (per cent)	92	94	95	95
Calves age at sale (days)	234	236	360	374
Calves weight at sale (kg)	257	266	341	379
Concentrates per cow (kg)	108	81	162	168
Concentrates per cow (£)	14	10	29	20
Concentrates per calf (kg)	25	15	184	203
Concentrates per calf (£)	3	2	21	24
Stocking rate (cows/ha)	1·5	1·9	1·2	1·3
Nitrogen fertiliser (kg/ha)	97	107	108	130

An average realisation of £347 for weaners sold from autumn-calving herds as opposed to £253 for spring-born weaners is impressive until one assesses the additional costs of obtaining this higher return; in particular the very much longer period that calves remain on farms (126 days on average), the greater cost of concentrates for both cow and calf (£33) and the increased area of land that is required for the autumn-calving herd (0·3 ha). The overall effect of these factors is that there is comparatively little difference in gross margin per hectare (£3/ha in favour of spring calving). When one looks at the performance of the top third of the herds for each group the margin in favour of spring-calving herds is £62/ha, an advantage of approximately 16 per cent.

It is true that the spring-calving farm will have a slightly higher capitalisation per ha because of the higher cow stocking intensity, but this will be offset by the extra capitalisation arising from holding weaners for an additional 18 weeks on average. The expectation is that autumn-calving farms will also have a higher investment in buildings, conservation equipment and conserved feed. The satisfaction that the autumn-calving farmer derives from his higher average realisations at the autumn-suckled calf sales can have a rather dubious foundation when all the relevant factors are taken into account.

Looking at timing of calving from a national viewpoint, there is no question if the above comparisons are valid for the whole sector of upland-suckled calf production, and there are no grounds for thinking otherwise—spring calving appears to be the better proposition because of its higher output of weaned calves per unit area. This will matter greatly if we move into a situation where there is a reduction in the number of dairy-bred calves available for beef production as a result of an enforced cut-back of the dairy cow population. If this is the prospect then it is important that a higher proportion of these weaners coming from the uplands, regardless of season of birth, are sired by bulls of the high growth beef breeds such as the Charolais. There is evidence that this is happening on both upland and true hill farms where once the traditional British beef breeds were dominant. High growth weaners are also in the interests of the individual producer because their superiority in this respect will earn a deserved premium in the sale ring.

FURTHER DEVELOPMENT OF UPLAND FARMING

The upland sector has scope for further development if the Community's appetite for beef and sheep meat is not fully satisfied, but it is doubtful if we shall ever again witness the type of transformation that took place over the 1950–70 period, even though there are still many thousands of hectares of hill and uplands that are by no means fully farmed. One argument against further investment is purely economic and it is not new as it was raised in the sixties when critics questioned the justification for massive injections of public money into the hill sector when it was only accounting for about 5 per cent of the gross product of British agriculture.

These critics took a rather too blinkered view in that they failed to give weight to two important elements; first the complementary contribution of upland farming to lowland productivity in the shape of store and breeding animals and, second, the amenity value of productive uplands. The majority of town-dwellers, who in this age of the car are able to visit and enjoy the more remote rural areas, are not in search of a neglected wilderness. It is part of their enjoyment to see a living countryside with farm stock and the people tending them going about their daily business.

More recently there has been another more strident expression of opposition to further land development in Britain. It comes from a relatively small but nevertheless vociferous and influential lobby representing conservation interests, which oppose any further developments in agriculture and forestry that could be at the expense of existing flora and fauna. Its views which are not limited to upland sites has greater force in the present situation where the EEC is in surplus for many of the farm products, in some instances to an embarrassing degree.

A full discussion of this contentious subject is not appropriate here, but it is pertinent to observe that during the war when the U-boat campaign was at its height there were few protests about the ploughing out of thorn and rabbit-infested downland or the traditional mowing meadows that produced less than a half crop of hay. The endangered species then was man, not butterflies or the weeds of ill-managed grassland that are euphemistically called wild-flowers by people who do not have to make their living from the land. It is important to retain a perspective on this matter and also to remember that the great upsurge of British agriculture, which was started and sustained by the exigencies of war, was a partly derelict and half-farmed countryside because of a neglect of agriculture from 1870 onwards by a nation whose preoccupation was with manufacture and trade. Many of the improvements that have been effected since 1940 have been little more than a restoration of land to its state in the period of high farming prior to 1870.

The social significance of reclamation in the upland areas of Britain should not be forgotten. If they had been left in their pre-war state or handed over completely to foresters there would have been a much greater exodus of people from this sector in the continuing process of rural depopulation. Farming people who remain have cause to be proud of their achievements. Apart from their contribution to the national larder they have added to the beauty of this countryside which now has an aura of modest prosperity that can also be enjoyed by townspeople. Farmers on such land are not in the habit of burning straw for they value it for its usefulness in their farming, and they are not given to bulldozing hedgerows. Their concern, generally is to plant more trees, preferably in wide belts to provide shelter for their stock and, in the process add further variety to the countryside.

Farmers realise very well that they have to share their land with the townsmen of these crowded islands and in most instances they are very happy to do so, provided gates are not left open, dogs are not permitted to worry sheep, and sundry items like old bicycle frames and clapped out TV sets are not thrown over the roadside dyke-backs. They will continue to produce beef and even expand its production if the country needs it, provided there are the necessary incentives to do so.

MARGINAL FARM CONTRIBUTIONS

Marginal land is defined as that category of less kindly land that is not naturally suited to intensive arable or livestock production and which is, for the most part, best maintained in permanent or long duration pastures. It is exemplified by the belt of strong land lying east of the Pennines in north Yorkshire, Durham and Northumberland and west of the coastal band of easier working land that is largely devoted to cash cropping or dairying. Much of the marginal land of the United Kingdom was in a very run-down state prior to 1940 because of a neglect of such basic measures as drainage, weed control, liming, fertilising and rabbit control. The cruel reality in those days was that there was insufficient income generated by these farms to pay for such inputs and farmers were in a vicious circle where low incomes perpetuated low productivity.

It is a class of land that is best farmed in moderately large holdings, not less than 200 ha, but preferably larger to give economies of scale and a reasonable income for the farmers involved. Its best use in normal circumstances is for breeding, both sheep and cattle, though sometimes smaller farms will carry dairy herds because of the higher income they will generate as compared with sheep or suckler cows. Some cropping can be undertaken but less for direct income than as a source of supplementary stock food. Often a forage crop such as rape or turnips will, apart from its food contribution, be a useful step in preparation of land for establishing a long duration pasture. Apart from lime, the most important fertiliser input is phosphate in order to get a good establishment of clover because of its vital role in pasture productivity. The main thrust in pasture management therefore, is the creation of balanced grass and clover swards and this means a more sparing use of nitrogenous fertilisers than that usually adopted on lowland grass. Its principal role, apart from stimulating early growth and sometimes foggage for back end grazing is one of augmenting pasture cuts for silage or hay. Hay was once the mainstay for winter feeding but increasingly it is being supplanted by silage as a much safer product. The overall aim is to make the best use of pasture, both grazed and conserved, for this is the cheapest and most appropriate basis for feeding in what is, relatively speaking, an insufficiently intensive system of livestock production to justify more than a tactical reliance on concentrates.

Marginal farms, generally, have been our most important sector in the production of suckled calves, and farmers specialising in this enterprise have now attained a very high state of expertise that will be discussed in greater detail in a subsequent chapter. As in the upland and hill sector, suckler production was encouraged at the expense of sheep production up until full membership of the EEC was attained. Now however, the balance has been restored with most marginal farms being situated in the so-called 'less favoured areas' and suckler cows qualifying for the Livestock Compensatory Allowance, which in 1981 was at the rate of £44·50 per cow. The expectation is that there will be a more stable and better balanced relationship between the two kinds of breeding stock on marginal farms.

This is a sensible situation from several viewpoints. First there is a better utilisation of pastures with a combination of sheep and cattle. Apart from the usefulness of sheep in utilising herbage that has little value for cattle, for example rank growth close to dung pats and residual grazing when cows move to their winter quarters, there is the important part that cattle can play on a mainly grassland farm in effecting a clean grazing policy for the sheep flock. Probably the most important advance in the intensification of sheep production since 1960 is the growing realisation that lamb performance is greatly enhanced when they have access to 'clean' grazing, that is pastures (or forage crops) that are free of infective larvae of the internal parasites that can play havoc with thrift. The use of cattle as a disinfecting agent, together with conservation as well as any maiden seeds that may be in the pipeline, gives the farmer the opportunity of planning clean grazing for his ewes and lambs so that a much higher proportion of the crop can be drafted fat off their mothers.

A formalised grazing programme could be of the following order: one-third for conservation, one-third grazed by ewes and lambs and one-third by cows and their calves. From November onwards ewes could graze on any pastures without appreciable detriment in respect of level of worm infection in the following season, while weaned lambs would go on to clean aftermaths only after they had been drenched to clear any level of infection they may have picked up, despite the precautions that have been taken. These aftermaths will also be very useful for the cows and calves for the latter will also benefit from a change of pasture.

We believe that a farm of this type should practise early spring rather than autumn calving for not only will this mean a better relationship between the availability of pasture and the nutritive needs of the stock but also it will make possible a higher stocking intensity. Marginal farms tend to be located on heavy land which can suffer from severe winter poaching and once pastures are put into reasonable heart most farmers find it advisable to provide some sort of inwintering accommodation. This must not involve heavy capitalisation for the income generated by a suckler cow, even one rearing a cross Charolais calf which is worth £350 or more as a weaner, does not justify anything grander than a simple set of cow kennels which have another advantage over covered yards in that they need a minimal amount of bedding.

It is important to have compact calving and early summer as opposed to the winter mating required for autumn calving helps in this respect. Not only does it mean that there will be a more even offering at the autumn-suckled calf sales but also there is no longer temptation to winter late calves for disposal in the spring. The aim on a farm like this should be one of wintering as many cows as possible, for winter is the bottleneck that determines more than any other factor the efficiency of pasture utilisation during the short growing season characteristic of such land. It should be a function of lowland and not marginal farms to winter stock intended for slaughter for they have the resources for this.

LOWLAND BEEF PRODUCTION

Only occasionally does one find suckled calf production on lowland farms, for such land, if it is to be used for beef production is better employed in finishing rather than breeding. Exceptions to this situation are encountered in arable districts such as parts of East Anglia where there are areas of grassland subject to winter flooding that provide useful summer grazing. Breeding herds of Lincoln Reds for instance are sometimes summered on such land but in the winter they return to their traditional role of utilising arable by-products and, in the process, treading straw into FYM which, despite the much greater use of artificial fertilisers, is still highly regarded by farmers growing potatoes and sugar beet.

Such instances apart, the main role of the lowlands should be one of finishing animals bred on the higher land or in dairy herds. First there is grass finishing either on long leys or permanent grass such as those found in Leicestershire, Northumberland and other parts of the United Kingdom. This is undertaken on land not suited to intensive arable production but nevertheless ideal for summer grass and clover production. More often beef production on mainly arable farms, with only limited grassland, usually a break in an arable sequence, is winter fattening in yards and the object of the exercise is again the production of farmyard manure as well as a profit from finishing.

On many arable farms with long leys and possibly some permanent grass the 18-month system of beef production, based on the rearing and finishing of dairy-bred calves, particularly those born in the autumn, is the most appropriate. MLC surveys reveal that it produces the highest and the most consistent gross margins per hectare of any system of beef production apart from barley beef where production per hectare is a meaningless statistic. The main drawbacks of the 18-month system are the high peak capitalisation and the relatively poor cash flow, especially when compared with yard or grass finishing of stores over a 3–6 month period.

Finally there is barley beef production which of course is best undertaken on arable farms that are producing the main ingredient of the diet. It and other forms of finishing beef cattle will be discussed in detail in subsequent chapters.

OTHER POSSIBLE SOURCES OF BEEF CATTLE

About 1960 the English Milk Marketing Board organised a field trial in dairy herds which aimed to increase multiple births by means of hormone treatments. On paper it promised to be the ideal solution of a national need, namely a large increase in the number of calves that could be reared for beef without any increase in cow numbers. Unfortunately it was not a success for a variety of reasons. In some cases fertilisation was delayed or was not achieved and this was to the detriment of milk production which is a dairy farmer's prime concern. In other cases there were 3 or more calves with an increase in neo-natal mortalities and here again milk yields can adversely be affected by multiple births. Where twins, the desired result, were obtained, in most instances where they were of the mixed sex variety the heifer was a free-martin and therefore incapable of breeding.

Looking with hind-sight at what was a highly commendable trial in some respects, it would have been preferable to use suckler cows because small upsets in milk yields would have been of lesser concern, while the free-martin condition would not be serious in animals intended for slaughter. However, it is clear that the state of knowledge of hormone manipulation was at that time insufficient to make this a practical proposition.

Possibly a more hopeful approach in effectively increasing the number of calves for rearing for beef, without an increase in cow numbers, would be through the implantation of suitable fertilised ova in deep-milking sucklers like the Hereford × Friesian, which have ample milk to rear 2 good calves. The techniques involved are now established and fairly well refined so the outstanding task will be largely one of organising a routine that is economically viable. It is estimated that a ewe rearing twins has at least 70 per cent advantage over a ewe rearing a single up to point of weaning, in terms of efficiency of food utilisation. Though the advantage for twinning in cattle would be somewhat less nevertheless it must still be substantial. Almost more important would be a better utilisation of the considerable capital investment that is tied up in a breeding herd. Altogether it is a more precise approach to multiple births than hormone manipulation, for at least there is little prospect of more than 3 calves being born at each parturition.

A less spectacular approach to increasing calf numbers without an increase in the number of breeding cows was initiated by the senior author at Cockle Park in the late sixties. At that time there were about a million maiden heifers that were being slaughtered annually in the United Kingdom and the question was asked whether it would be feasible to mate a proportion of these at a relatively early age to produce a calf and a carcass before they were 27 months of age. Heifers can be very precocious in respect of sexual maturity if they are reared on a high plane of nutrition. For instance it is not uncommon in suckler herds for heifers of only 6–7 months 'to steal the bull' when it is left with the herd to cover late calving cows. The first trial at Cockle Park was much less ambitious in respect of age at mating and a dozen heifers, some of them dairy-bred and some out of sucklers, were sufficiently well grown for successful mating to an Aberdeen Angus at 12 months of age. They all calved successfully at 21–22 months and they suckled their calves for 3 months. The heifers were then given a finishing diet of silage with a barley-based supplement and they were in slaughter condition in less than 3 months, because they very quickly restored the flesh they had lost during suckling.

Cut-up tests revealed that there were no apparent harmful effects of carcass proportions and a master butcher catering for a select trade had no complaints about quality from his viewpoint, while expert panel tests gave the beef a higher rating than that from well-finished Irish bullocks. We estimated that these heifers produced 36 kg more carcass than they would have done at a normal slaughter age of 18 months. In addition they produced useful calves and gave them a good start in life.

We used Aberdeen Angus semen in order to reduce birth weights and the risk of difficult calvings. In this we were successful but naturally, there was some loss of size in the resulting calves. There should be no great danger from using breeds like the Sussex and

Hereford which have reasonably good records in respect of dystocia provided the plane of nutrition is lowered about a month prior to calving. Medium grades of spring-born suckler heifers would fit in very well with this system, which we have dubbed 'bred-heifer' production, because they will be sufficiently well grown, following fairly cheap wintering, to be mated at the yearling stage while it is unlikely that they would reach slaughter condition before the end of their second summer if they were not bred. Autumn-born heifers on the other hand will normally be in slaughter condition during their second winter at about 15 months of age and it is unlikely that it would be profitable to retain them for another 12 months before slaughter. Dairy-bred heifers, in particular Hereford × Friesians, are well suited to the system because it is relatively easy to effect mating at the 12-month stage. Moreover they have sufficient milk to give their calves a good start in life.

Though the Cockle Park trials were successful and though Irish workers got similar encouraging results, the bred-heifer did not become a commercial proposition in Britain. It was not a question of shortcomings in the concept but limitations enforced by the prevailing regulations for deficiency payments which stipulated that heifers that had carried a calf, even though they had no more than 2 broad teeth, were not eligible for the payments that supported the guaranteed price.

The situation could change quite dramatically if there is a deliberate cutting down of dairy cow numbers in the EEC in order to limit milk production. Then the additional calves that bred-heifers could provide might become so important that they would deserve positive encouragements instead of the bureaucratic discouragement that the bred-heifer concept first experienced.

Chapter 3

BEEF BREEDS AND CROSSES

THE GREAT pioneer improver of British livestock was a Leicester-shire farmer, Robert Bakewell (1725–95). He was more successful with sheep than with cattle; the Dishley Leicester, which was the forerunner of the modern English Leicester and one of the parent breeds of Border Leicester, also contributed to the improvement of all the Longwool breeds. His choice in cattle, the Longhorn, was not a happy one; though he made considerable progress in improving beefing qualities, its horns were a serious drawback with the development of yard finishing of cattle, an integral part of the Norfolk four-course rotation that spread widely through Britain in the wake of the enclosure movement. Cattle, then, were not just a profitable means of utilising the clover and turnips, which were key crops in the rotation, but they were also important agents in treading straw into dung which was essential for maintaining soil fertility.

EARLY DEVELOPMENTS

Bakewell was a remarkable man whose ideas in breeding were in advance of contemporary science and have only been substantiated by later advances in knowledge. He lived and worked at a critical time in the development of British agriculture. The creation of enclosed farms had made it possible for the yeoman farmer to make independent decisions about the breeding and feeding of his livestock, and turnips and clover allowed him to raise the plane of feeding so that there could be an expression of genetic differences in important traits such as growth rate, depth of flesh, and body proportions.

Bakewell attracted many visitors to his farm at Dishley, among them Robert Colling, who with his brother Charles worked with a different foundation, the Teeswater and Holderness cattle, which had among other things some admixture of Dutch blood that had been introduced after the Restoration. The Collings bred the bull, *Favourite*, which, by all accounts, was a quite outstanding animal whose breeding qualities put the Shorthorn breed, as it came to be known, on the map. Others followed in the Collings' footsteps, notably the Booth family, father, sons and grandsons, who for a period of over a hundred years covering virtually the whole of the nineteenth century selected for fleshing qualities. Meanwhile another master breeder, Thomas Bates of Kirklevington selected for dual-purpose functions. He developed the famous Duchess strain of Shorthorns which was eventually lost through sterility presumably due to inbreeding, which was an important tool with early breeders in their endeavours to ensure purity of descent from favoured ancestors and with it uniformity of type. Figure 1 illustrates this point and this pedigree is by no means atypical of those created by the early improvers.

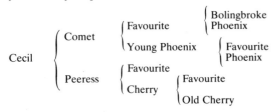

Fig. 1. The Collings' intensive use of the bull, Favourite, *as a preferred ancestor (Kelly, 1946).*

Early on, Scottish farmers became interested in the Shorthorn and one of these, Amos Cruickshank, bought without first seeing it an English-bred bull called *Lancaster Comet*; he was so disgusted with it on its arrival that he turned it out with some shy breeding cows on a remote farm. *Comet* did not survive the rigours of a Scottish winter but he left behind some calves including one that was aptly named *Champion of England*. Like *Favourite* he left a mark behind him and Cruickshank's famous herd was largely descended from this bull. Cruickshank defined as his breeding objective cattle that would finish on turnips and oat straw, and in selecting to this end he developed compact, early-maturing cattle with a great depth of flesh, that became known as the Scottish Shorthorn.

Other Scots were working with local cattle and the most famous of these breeders were Hugh Watson of Keillor and McCombie of Tillyfour. Before Watson started his breeding work the cattle of Angus and adjacent counties were a fairly motley lot, with some horned and others polled, and a range of colours including duns and reds and blacks. He opted for black polled cattle and so did McCombie. Evidently the latter had trouble with red recessives, because it is reported that he painted his buildings black and kept his breeding cows in walled enclosures to avoid undesirable maternal impressions. One can smile at such naivety today, but there is no denying the remarkable attainments of these early breeders who laid the foundations of the Aberdeen Angus breed which has been recognised the world over for the quality of its beef.

Elsewhere in Britain there were pioneers working on other local races of stock and though space does not allow an account of all of them, mention must be made of the Tomkins family, which laid the foundations of the modern Hereford. Initially, like so many British breeds, Hereford cattle were valued for draught purposes, but even in the eighteenth century cattle from Hereford had a reputation as good grazing animals. Again the early breeders selected for uniformity, especially in respect of characteristic markings, depth of fleshing and early maturity.

BRITISH BREEDS IN THE NEW WORLD

These three breeds, above all others developed in Britain, achieved much more than a national reputation. With the opening up of the New World, British settlers took with them their favoured breeds of stock which were quickly adopted by settlers coming from other European countries. Up till comparatively recently the only breeds of cattle and sheep of any consequence in North and South America, South Africa and Australasia that were not of British origin, were Merino sheep and Friesian (Holstein) and Brown Swiss dairy cattle. In Argentina the Shorthorn was the dominant beef breed until the early nineteen-sixties, followed by the Aberdeen Angus and Hereford. All three breeds and especially the Hereford are important in North America, away from the higher temperature regions where stock of Zebu origin are now achieving greater importance. Some of the new breeds that have been evolved for these regions contain a considerable proportion of British blood, notably the Brangus and the Santa Gertrudis which combines Shorthorn and Brahman genes.

Why have British stock and especially the beef breeds gained such an ascendancy in these new countries? Probably the simple answer is that in the nineteenth century these were the most improved animals available, because British breeds, both cattle and sheep, were also introduced into Continental Europe to act as improvers of local stock. For instance, it has been stated that the Shorthorn has contributed to the Charolais. Unquestionably the early development of enclosures gave Britain a head start in the improvement of grazing animals, both in the control of stock for breeding purposes and in the size of herds and flocks which gave individual farmers scope for selection. Then there are inherent qualities of stockmanship, which are particularly well developed in

Scots, and this capacity to present animals in first-class condition still remains and is an important factor in their sale.

The capacity of the principal British beef breeds to fatten on low energy diets has also been an important factor, because animals that can finish on straw and turnips are well suited to pastoral conditions such as those in Argentina, Australia and New Zealand. With the development of an international trade in refrigerated beef, principally directed towards the British market, it became important to produce a carcass with a reasonable fat cover; this not only improved eating qualities but preserved a bloom on carcasses in storage and transit. Gradually this trade moved more and more to a demand for small tender joints, which meant the use of early-maturing animals that could be slaughtered at a comparatively early age with the right degree of finish. The Shorthorn and the Aberdeen Angus in particular suited this demand.

This in its turn had its effects on breeding objectives in Britain and many of the leading breeders, especially Shorthorn and Aberdeen Angus, and to a lesser extent Hereford breeders, made their primary goal what has come to be known as the 'export type', which is a very compact, short-legged and deeply fleshed animal. Rate of growth and ultimate size were not considered to be particularly important. In fact the preference appeared to be away from heavy mature weights. This concern with the 'export type' was stimulated by overseas purchasers who attended the top British sales and especially those at Perth. Argentinian ranchers, while their country's overseas funds permitted this, were high bidders at these auctions and they had competition from other countries including the United States, which also had breeders who were obsessed with the compact type.

In fact some American breeders got themselves into trouble with 'snorter dwarfs' in the Hereford. However, they did see the light and set about changing the concept of what they sought, aiming at a longer legged animal more suitable for range conditions. Fortunately for the UK, Hereford breeders did not suffer the same problems with regard to 'snorter dwarfs' and there has always been a substantial proportion of larger-framed animals for crossing purposes.

CHANGING DEMANDS IN BRITAIN

Out of all this arose an ironical situation, that crossing bulls of our three most important beef breeds were a by-product of breeding for the export type in herds at the top of the breeding pyramid. This did not matter so much nationally before the war, when there was cheap imported beef to augment home production that was mainly a by-product of dairying, but today with the United Kingdom very reliant on its own production, including that from more than a million specialised beef cows, there can no longer be a neglect of rate of gain or of weight at slaughter. In consequence a different pattern of breed demand is emerging. The Beef Shorthorn from its once proud position is now a very minor breed; there were only 272 bulls licensed in 1972 as compared with 5,457 Herefords.

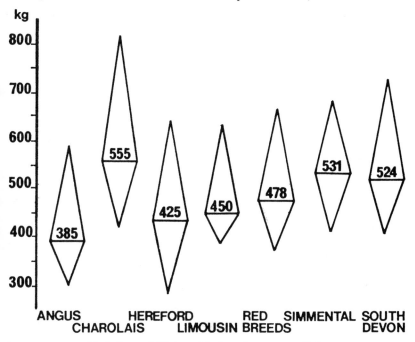

Fig. 2. Range of 400-day weights in performance-tested bulls.

It is most improbable that the Shorthorn will ever become important again because it has been bred so far into the ground.

Similar fears have been expressed for the Aberdeen Angus, but as the data in Figure 2 show, there is still a considerable range in 400-day weights for the bulls that have been performance tested by the Meat and Livestock Commission; unquestionably there is still sufficient genetic variability for breeders to exploit should they elect to do so. The breed is at a cross-roads on both sides of the Border though further north in Scotland there are still areas where there is a lot of loyalty for their native breed, partly on the grounds that it has no equal in respect of meat quality and therefore of basic importance in sustaining the reputation of Scottish beef. Less sentimental farmers, conscious of the importance of growth performance, have opted for other breeds. First it was the Hereford but it was soon superseded in suckler herds by the Charolais and other Continental breeds once their merits were appreciated and the price of the bulls became commercially acceptable.

Both Aberdeen Angus- and Hereford-cross cattle are good fatteners on pastures or other high roughage diets. Pure Friesians or their crosses with the Charolais or Simmental tend to keep on growing, rather than putting on the necessary finish when on pasture and it is only with a fair measure of maturity that they will achieve the necessary finish. The same is true with a diet consisting largely of silage and unquestionably the premium that Hereford-cross Friesian bull calves enjoy over the pure Friesian, especially away from the arable districts, is largely attributable to their capacity to finish either with low levels of cereal supplementation or even with none at all. The propensity for easy fattening is something of a disability in heifers which are much earlier maturing than steers, because they can lack sufficient size at slaughter. Aberdeen Angus and Hereford breeders must strike a balance between size and rate of growth on the one hand and carcass attributes on the other if their respective breeds are to retain a major role in beef production.

CHANGES IN BREED PREFERENCES

Since the early sixties there has been a remarkable interest in the United Kingdom as well as overseas in breeds other than the Shorthorn, Aberdeen Angus and the Hereford that once dominated specialised beef production. Not only did this lead to the importation of several breeds, mainly from the Continent, but also a new look at some of the other British breeds which previously had little more than local importance. Some imported breeds have been ships that passed in the night but there are at least three that now have a firm foothold in this country, namely the Charolais, Simmental, Limousin and possibly a fourth in the Blonde d'Aquitaine. Unquestionably the starting point in this interest in new breeds was a growing realisation that traditional pedigree breeders were not supplying the sort of bulls that met the needs of commercial beef producers. This led to the importation of a trial shipment of Charolais bulls which were limited in their use to artificial insemination until their worth was assessed. The breed came through with flying colours and the next step was the importation of females and the establishment of pedigree herds. The Charolais was soon followed by other breeds but some like the very big

Italian breeds had very little to recommend them and were soon discarded.

The Beef Recording Association, established in 1964, contributed to the interest in less-fashionable breeds because it collected information on the performance of different breeds and crosses over a range of production conditions and, for the first time in Britain, facts instead of subjective appraisals became the basis for comparisons between the different kinds of cattle. The BRA was also responsible for initiating a programme of performance testing for the major beef breeds where young bulls were compared for growth rate under standard conditions. With the setting up of the Meat and Livestock Commission the BRA was disbanded and its work was taken over and greatly extended by this new body. On-farm recording of pedigree herds is one of its important functions and Table 3.1 gives details of herds recorded in 1982.

TABLE 3.1. HERDS IN MLC PEDIGREE RECORDING SCHEME, 1982

Breed	No. of herds	No. of cows	Av. cows/herd
Aberdeen Angus	63	1,633	25·9
Blonde d'Aquitaine	23	233	10·1
Charolais	389	4,035	10·4
Devon	17	575	33·8
Hereford	166	4,870	29·3
Limousin	159	1,371	8·6
Lincoln Red	14	912	65·1
Simmental	246	2,100	8·5
South Devon	57	1,449	25·4
Sussex	23	862	37·5
Welsh Black	20	960	48·0
Total	1,177	19,000	16·1

There are several absentees from this list, including the Galloway and Luing but this is understandable for these are breeds of difficult environments where recording is not easily undertaken. Another absentee in 1982 was the Beef Shorthorn but this is a reflection of its reduced status. One interesting feature is that no less than two-thirds of all recorded herds are accounted for by the 4 imported breeds with the Charolais in first place with a one-third representation. This indicates the progressive approach of farmers who have these new breeds with that of many Aberdeen Angus and Hereford breeders who are sometimes inclined to disparage 'paper' breeding and believe in a divine-given talent, described as 'breeder's eye'. Initially it took a lot of persuading to get these two breed societies to take part in bull performance testing but the growing popularity of the Charolais was an effective persuader. There has been a rather different attitude among those who are concerned with the future of less-fashionable British breeds. The Lincoln Red Society was the first to organise on-farm recording, some years before there was any national body with this responsibility, and the other red breeds (the 2 Devons and the Sussex) have been well represented in recording and performance testing.

A notable feature of Table 3.1 is that most recorded beef herds are small. The outstanding exception in this respect is the Lincoln Red with an average breeding herd of 65 cows contrasting with 9s and 10s for the Continental breeds. The latter, of course, are in the early years of establishment and there has not yet been time for them to multiply to the point where individual breeders can afford to have herds of the order of 40 or more breeding females. It is only with herds of this strength that contemporary comparisons based on farm records become reasonably informative. When the effects of sex, age of dam and season of birth are taken into account, and these are substantial in respect of weight for age, the scope for meaningful contemporary comparisons is very small. There are possibly no more than 50 breeders among 1,900 represented in Table 3.1 who are able effectively to utilise their records in formulating selection plans. This is where breeders in the newer countries, particularly in Argentina, USA and the older British Dominions, have a very considerable advantage because herd sizes are much greater and contemporary comparisons have real validity for the progressive breeder who is prepared to use them.

Though most of the participants in on-farm recording may derive little apart from general interest from the records they keep, nevertheless collecting these give a very informative picture of the average growth performance of the breeds over a range of environmental conditions. Average 200-, 300-, 400- and 500-day weights for heifers and bulls in MLC-recorded herds over the two-year period 1981–83 are summarised in Table 3.2.

Heifer weights are generally much more reliable than bull weights as indicators of breed's weight-for-age performance because there is greater likelihood of the latter being distorted by preferential feeding to boost their sale value. The amount of such feeding varies between breeds and also within herds. When Perth sales featuring Aberdeen Angus or Shorthorn bulls were major annual events attracting large numbers of buyers, including contingents from overseas who were prepared to pay high prices for bulls that took their fancy, the use of the nurse cows was widespread. Indeed milking ability in these 2 breeds was so neglected that in countries such as Australia and Argentina, where commercial cows are expected to rear good calves, this neglect became

TABLE 3.2. AVERAGE BEEF BREED WEIGHTS 1981–83 (KG)

Age in days	Bulls				Heifers			
	200	300	400	500	200	300	400	500
Aberdeen Angus	218	311	407	491	190	250	300	351
Beef Shorthorn	239	325	416	524	192	244	308	317
Blonde d'Aquitaine	261	379	501	599	227	308	377	462
Charolais	299	432	565	677	260	349	421	482
Devon	204	299	429	552	181	240	289	325
Hereford	224	325	430	527	194	258	315	364
Limousin	258	376	500	604	229	306	378	424
Lincoln Red	248	362	500	na	209	266	317	366
Simmental	294	421	553	655	252	330	397	457
South Devon	287	411	544	653	236	309	370	425
Sussex	226	428	437	522	200	270	329	392
Welsh Black	234	331	441	na	205	252	300	353

a breed detriment. Certainly the massive decline that occurred in the popularity of the Shorthorn breed in Argentina after 1955 is attributable in some measure to a deterioration in milking ability as well as growth performance. The use of nurse cows was very rare in the unfashionable breeds like the Sussex, Lincoln Red and Devon which are expected to rear their own calves. The milking propensity of both the South Devon and the Welsh Black has never been in question for up until quite recently they were generally considered to be dual-purpose breeds. This is also true of some of the Continental breeds in their native localities, but in Britain they are now considered to be specialised beef breeds.

The most striking feature of Table 3.2 is the weight superiority of the imported breeds over the British breeds, apart from the South Devon which is the largest of the native breeds. There is little difference between it and the Simmental but the Charolais has a small weight advantage over both. The relatively small size of the Welsh Black is not a disadvantage for its role is not primarily one of providing sires for terminal breeding but females for single suckling herds, especially under upland conditions. Its worth in this respect is now recognised outside of Wales. Another breed, not included in Table 3.2, is the Galloway which has probably a better adaptation to difficult environmental conditions than any other British breed, with the possible exception of the Luing. The latter is a synthetic breed of comparatively recent origin which was developed by the Cadzow family in Scotland from a Beef Shorthorn–West Highland base. Females of this cross, which is known as the cross Highland, have a long-established reputation as good commercial sucklers on hill farms and in this respect they compare with the better known Blue Greys which are the produce of White

Shorthorn bulls on Galloway cows. There is not a lot of information available on the Luing which is still a fairly minor breed but it is known that the Cadzows have been very realistic in their selection programme, which is directed to the development of cattle that are well suited to the high rainfall uplands of the West of Scotland. This is, of course, the home of the West Highland, that picturesque breed with a long shaggy hair and an extravagant spread of horns which is now a very minor breed.

PERFORMANCE TESTING OF BEEF BREEDS

It is fortunate that weight-for-age is a strongly inherited characteristic for this means that the weight performance of the parent is a fairly reliable indicator of its breeding worth in respect of this trait. It has other merits for it can be measured with reasonable accuracy in the live animal during rearing so there is no undue extension of the generation interval, which is one of the great detriments of progeny testing, while measurement is also a relatively cheap operation. Performance testing was offered as a service to breeders by the Beef Recording Association soon after its establishment and this was maintained and extended by MLC and the figures on which Figure 3.2 is based are the fruits of this work. Currently there are 5 performance testing stations in operation and between 400 and 500 young bulls are tested annually under standard management.

Table 3.3 presents a summary of central performance test results for 1982–83 when 8 breeds participated including yet another imported breed, the Australian Murray Grey. Its performance however does not suggest that it will increase in popularity. In particular, its high backfat, a measurement that is taken at 12 months of age, is a considerable detriment. This is also a shortcoming of the Hereford but the range of this measurement suggests that there is considerable scope for selection if breeders are

TABLE 3.3. MLC PERFORMANCE TEST RESULTS 1982–83

| Breed | No | 400-day weight (kg) | | Backfat (mm) | |
		Average	Range	Average	Range
Blonde d'Aquitaine	13	580	523–634	1·9	1·6–2·4
Charolais	71	641	552–734	2·3	1·4–3·6
Hereford	51	510	436–572	4·8	2·4–8·0
Limousin	66	545	420–626	2·1	1·3–3·2
Murray Grey	10	477	408–570	5·5	4·1–7·4
Simmental	136	628	542–740	2·6	1·5–4·1
Sussex	18	518	478–600	3·2	2·2–4·9
Welsh Black	15	540	450–628	2·5	1·6–3·5

inclined to make the effort to improve the breed in this respect. All 4 Continental breeds are characterised by low backfat measurements and here again this gives grounds for the growing popularity of these breeds, particularly the Limousin, with the meat trade.

The high 400-day weights achieved by the Continental breeds is but further confirmation of their general superiority over British breeds in rate of growth, while the numerical strength of these breeds again indicates the progressive attitude of the breeders concerned. Because they are mainly working with a very few animals, central performance testing is of particular value to them as compared with breeders who have many contemporaries in their herds.

PERFORMANCE OF CROSS-BRED SUCKLER COWS

A very high proportion of specialised beef production is dependent on cross-bred animals. For instance most suckler herds have cross-bred cows with Blue Grey and the Hereford-Friesian as the favoured crosses. It is also customary to mate these cross-breds to a bull of a third breed to produce a two-way cross slaughter generation. There are sound biological and economic reasons for this reliance on cross-breeding, particularly the greater adaptability and reliability of the cross-bred cows as compared with pure-breds attributed to hybrid vigour and the contributions that come from compensatory mating, particularly in respect of growth rates and carcass quality in slaughter animals. Obviously there must be pure breeding in order to maintain the individual breeds and some of these, notably the Galloway are particularly well suited to difficult ecological situations. One thinks however, even with pure Galloway herds less in terms of breeding more pure breds (apart from the minimum required for herd maintenance) but more as a source of Blue Grey heifers which are so highly regarded by suckler calf producers in less favourable farming environments. They provide one of the few instances where heifer calves are more valuable to the farmer than bull calves, because of the premium the heifers will subsequently earn at disposal.

SIRE PERFORMANCE BY BREEDS

With most farmers a prime criterion in suckled calf production is weight for age for this is the main determinant of price but there are other important considerations as well, as Table 3.4 based on MLC data reveals.

The Charolais' bad record in respect of assisted calvings is almost matched by the Simmental and the South Devon. All 3 breeds have more than double the rate of assisted calvings of cows

TABLE 3.4. COW PRODUCTIVITY IN RELATION TO BREED OF MALE

Sire breed	Assisted calvings %	Calf mortality %	Calving interval days	Annual calf prodn/cow kg
Charolais	9·0	4·8	374	208
Simmental	8·9	4·2	374	203
South Devon	8·7	4·0	375	203
Devon	6·4	2·6	373	200
Limousin	7·4	3·8	375	199
Lincoln Red	6·7	2·0	373	198
Sussex	4·5	1·5	372	196
Hereford	4·0	1·6	372	189
Aberdeen Angus	2·4	1·3	370	179

mated to the Hereford, while the Aberdeen Angus is even better with a figure of only 2·4 per cent. As expected calf mortality rates are closely related to those for assisted calvings. These 2 sets of figures combine to underline the wisdom of using either the Angus or the Hereford rather than any of the heavier breeds on maiden heifers and also, as many farmers have learned to their cost, the danger of getting cows mated to the big breeds in high condition prior to calving. This is a point which will be discussed in detail in Chapter 12.

Choice of sire does not appear to have much effect on calving interval though the data suggest that there is a small advantage in favour of the lighter breeds. It would be interesting to have this measurement over several years to determine whether there is any cumulative effect, for there is a suspicion with some farmers that the continued use of bulls of the larger breeds is at the expense of longevity and lifetime performance. However, on the other side of the ledger there is a very clear advantage in favour of the bigger breeds in respect of annual calf production per cow.

ENVIRONMENT AND BREED PERFORMANCE

Table 3.5 gives a summary of sire performance according to breed on lowland, upland and hill farms.

The several breeds have practically the same ranking under each of the 3 farming environments and there can be no suggestion from the data that there is an interaction between sire breed and environment, except that the Hereford appears to have a consistently poorer comparative performance on upland farms than it has on either lowland or hill farms. No explanation can be advanced for this apart from possible sampling idiosyncrasies, for there has never been any well-founded view that the Hereford performs less well under upland conditions than it does on lowland farms.

TABLE 3.5. EFFECT OF SIRE BREED ON 200-DAY CALF WEIGHT (KG)

Sire breed	Lowland farms	Upland farms	Hill farms
Hereford	208	194	184
	Difference from Hereford cross (kg)		
Charolais	+ 32	+ 33	+ 21
Simmental	+ 24	+ 28	+ 14
South Devon	+ 23	+ 27	+ 16
Devon	+ 17	+ 21	+ 7
Lincoln Red	+ 14	+ 20	+ 5
Sussex	+ 7	+ 13	+ 2
Limousin	+ 7	+ 10	+ 2
Aberdeen Angus	− 14	− 12	− 8

In general the several breeds have a similar ranking for the 200-day weights of their cross-bred progency as that of the pure-breds. (See Table 3.2.) There are, however, two exceptions in the Devon and the Sussex. As pure-breds they are very similar in weight to the Hereford but both, particularly the Devon, have consistently heavier cross-bred calves than those sired by the Hereford. The Devon's cross breeding performance also compares more than favourably with that of the Limousin, which as a pure-bred, has higher 200-day weights. The explanation for this might be a more liberal level of supplementary feeding of the pure Limousin calves prior to weaning as compared with the Devons which are probably kept under more work-a-day conditions.

POST-WEANING PERFORMANCE ACCORDING TO SIRE BREEDS

Results obtained in MLC feeding trials with both suckler-bred and dairy-bred steers reflect their breeding in respect of rates of maturity and liveweight gain. Steers sired by bulls of late-maturing breeds are the fastest growing animals but they also need a longer feeding period to reach a given degree of fatness. This assumes the animals are all on the same diet but in commercial practice a farmer is able to accelerate maturity in late-maturing breeds by increasing the energy content of the diet. Table 3.6 is a summary of results for winter-fattened suckler steers, representative of 9 of the sire breeds that have been on recorded intakes to enable comparisons in economy of food use. Steers have been slaughtered at fat class 4L (determined by ultrasonic measurement), which is a very typical carcass grading. On the whole, steers sired by British breeds are slower growing than the progeny of the 3 Continental breeds, but the Lincoln Red is a notable exception to this generalisation.

Table 3.6. Performance of Winter-fattened Suckler Steers (Slaughtered at Fat Class 4L)

Sire breed	Daily gain (kg)	Feeding period (days)	Total feed (kg DM)	Kg feed/ kg gain	Slaughter wt (kg)
Aberdeen Angus	0·77	105	826	10·3	393
Charolais	0·84	148	1354	10·9	494
Devon	0·78	122	975	10·1	419
Hereford	0·78	120	950	10·1	410
Limousin	0·78	145	1201	10·5	454
Lincoln Red	0·85	117	999	10·0	428
Simmental	0·86	145	1324	10·6	490
South Devon	0·77	145	1196	10·6	451
Sussex	0·76	140	1118	10·5	428

It has also the lowest intake of dry matter per kg of liveweight gain, but gut-fill may account for these apparent advantages as the killing-out percentage figures in Table 3.7 reveal. Data in Table 3.6 suggest that a Lincoln Red-cross is possibly a more useful farmers' beast than a Limousin-cross but these reveal only part of the story for the Limousin has a 2·4 advantage in its killing-out percentage, a 2·5 advantage in percentage of saleable meat and a 1·1 per cent advantage in best cuts. It is not surprising that the Limousin has achieved a reputation for being an outstanding butcher's beast. The Charolais comes second in this appraisal with a killing-out percentage of 54·8, 72·7 per cent of saleable carcass meat and 44·8 per cent of best cuts.

Table 3.7. Carcass Results for Winter-finished Suckler Steers

Sire breed	Killing out (%)	Carcass wt (kg)	Saleable meat In carcass (%)	Best cuts (%)
Aberdeen Angus	52·5	205	72·5	44·1
Charolais	54·8	268	72·7	44·8
Devon	52·7	219	71·6	44·0
Hereford	52·3	214	71·9	44·1
Limousin	54·7	247	73·3	45·4
Lincoln Red	52·3	222	70·8	44·3
Simmental	53·0	258	72·0	44·8
South Devon	53·2	237	72·0	44·3
Sussex	53·1	226	72·6	43·9

MLC results relating to beef sires in dairy-beef production give more or less the same picture as that for suckled calves. The Continental breeds, in particular the Charolais, excel in respect of growth rate, killing-out percentage, carcass score and percentage of high-priced cuts. As with the sucklers, the progeny of the big, late-maturing breeds require more food than the progeny of the earlier-maturing breeds to reach slaughter condition on a given

Northeastern Junior College
LEARNING RESOURCE CENTER
STERLNG, CO 80751 57390

feeding regime. However, when this is related to carcass weights there is no meaningful difference in economy of food use.

Dairy-bred calves are the basis of cereal-beef production and here the Charolais, Simmental and South Devon crosses with the Friesian are unquestionably the best material because they do not become excessively fat before they reach reasonably acceptable carcass weights. This is the great shortcoming of calves sired by either the Hereford or the Aberdeen Angus. However, as pointed out earlier in this Chapter the situation is reversed with grass finishing for here Angus- and Hereford-crosses will achieve reasonable carcass weights and finish at the 18-month stage on a diet of pasture alone. The pure Friesian, or Charolais or Simmental cross Friesian steers though are unlikely to finish before 18 months on a diet of pasture alone, necessitating either cereal supplementation or later finishing in yards. There are indeed horses for courses and the British beef industry has no shortage of runners in this respect.

Chapter 4

BREEDING AND SELECTION (1)

Basic Principles

VARIATION IS a characteristic of any livestock population, no matter how conscientiously one attempts to breed for uniformity or endeavours to give all animals equal treatments. We can partition this into two main components, genetic and environmental variation. There is a third component which arises from the interaction between genotype and environment which contributes to adaptation. We have already considered examples of this, for instance, the greater fattening propensity that early-maturing breeds have over late-maturing breeds when they are being finished on pasture and high roughage diets.

GENETIC VARIATION

Genetic variation in its turn has two principal components which are additive genetic variance, where many genetic factors contribute to the expression of a trait, and variation which is due to the interaction of the genetic factors or genes. For instance, an animal may be very good because it has more than an average complement of genes that have an additive effect on expression of important traits, and equally one can have the reverse situation. The phenomenon known as hybrid vigour illustrates the effects of gene interaction. Perhaps the best examples of this are to be found with plants, particularly hybrid maize where parent lines have been selected not for their own individual production characteristics but for their performance in combinations with other lines. A similar approach has been used in the development of hybrid poultry, while with larger livestock, farmers have for a long time recognised the advantages of using certain first crosses, especially in respect of reproductive performance. For instance, Dr H. P. Donald of the Animal Breeding Research Organisation, has shown an appreciable improvement in lambing performance of first-cross Swaledale-Blackface ewes over the average of the two parent breeds.

The breeder of purebred beef cattle can do little about achieving greater genetic merit through gene interactions, except for any that might arise through an outcross. His concern, by the nature of his task and the resources at his disposal, must primarily be with additive genetic variance, so that he is able to supply his customers with bulls that transmit as high a complement as possible of desired genes that determine the expression of economic traits. He will leave it to his farmer customers to decide whether these bulls are used for cross breeding, either in compensatory matings or to exploit the possibilities of hybrid vigour.

The early breeders, who developed the distinct breeds that we now have, were concerned not only with the quantitative traits, that are an expression of many genes acting in concert, but also with qualitative characters such as colour, colour patterns and presence or absence of horns. These, for the most part, are determined by one or two pairs of genes and therefore constitute relatively simple genetic situations. Probably McCombie and Watson in their endeavours to evolve cattle that were true breeding polled blacks felt differently about their problems because at that time there was no established explanation of the mechanisms of inheritance. It so happened that their policy of breeding closely to favoured ancestors helped them in their task of achieving genetic purity for these characters.

POLLED CATTLE

Now that a reasonably high level of genetic purity has been achieved within breeds for the qualitative characters, the emphasis very rightly has moved to the quantitative characters which have greater economic significance. A principal exception is where breeders may be endeavouring to produce polled cattle in a breed that is normally horned. This is the present situation in Herefords which are of the two kinds, horned and polled. The polled cattle can be of two kinds as well, homozygous polled cattle that carry the two dominant genes for polledness (PP) and heterozygous polled (Pp) with only one gene for the polled condition that prevents the growth of horns.

The homozygous animal will always breed polled offspring but if heterozygotes are mated there is, over large numbers, the following ratio, 1 PP : 2 Pp : 1 pp. Once a parent has produced a horned calf then one can be certain that it is a heterozygote. However, a homozygous polled animal will always produce polled calves even from horned mates. This in fact is the best method a breeder has of determining whether a bull is a heterozygote or not. Ten polled calves from ten horned cows is as certain an indication of homozygosity as ten successive heads is an indication that a double-headed coin is being used in a tossing game. Although polledness is mainly a dominant character, there are instances where it can be recessive in certain breeds. Such a situation exists in the British Friesian where horned stock can occasionally give rise to polled offspring.

When one is dealing with a recessive character, in which the parents appear normal and the offspring are not, the breeder's task is more difficult. He can test mate his bulls by mating them to known carriers but this is obviously expensive and time-consuming. Genetic action against recessive traits is inevitably a lengthy process and often one is better to cut one's losses and eliminate animals whose immediate ancestors are known carriers. But always one must evaluate the potential damage likely from the defect against the potential value of the animal in question. Among such characters are things like bulldog calves and dwarfism.

STRENGTH OF INHERITANCE

The problem facing a breeder when he selects a herd sire is one of knowing how much of the apparent or phenotypic excellence of the animal he is considering is attributable to additive genetic factors and how much is due to fortuitously favourable gene combinations or to better than average environmental effects. There is no means of assessing this in the individual animal except by progeny test, and this in cattle is a long and expensive operation. It involves committing a considerable number of females, certainly not less than 20 but preferably more, and if the bull proves to be well below standard this can involve a considerable financial loss. Equally if the bull is a very good one the breeder is well along the way to a full exploitation of this genetic merit, and so it is important for him to take such steps as he can to improve the chances of selecting a good bull.

This leads us to a consideration of heritability or strength of inheritance of traits with a polygenic basis, such as milk production, growth rate and economy of food conversion. Population geneticists have been able through the study of records of close relatives to determine the importance of additive genetic variance, relative to the total phenotypic variance in breeding groups. Preston and Willis in their authoritative book, 'Intensive Beef Production' present a comprehensive summary of the estimates that have been obtained from a range of breeds under different environmental conditions, for the principal characters that are of importance to beef producers, and anyone interested in a deeper study of this topic is advised to consult this book.*

* Preston, T. R. and Willis, M. B. (1970), *Intensive Beef Production*, Pergamon Press, Oxford.

These estimates show a wide variation, not only from character to character, but also within characters, and this is understandable. For instance, characters that have been subject to intense selection, either natural or artificial, over a long period of time, will generally have very low heritability values; this is the case with most of the reproductive traits such as fertility, calf survival and conception rates. Within traits, breeding groups that have been subjected to intense inbreeding, have lower heritability values than those for groups with a low coefficient of inbreeding. Obviously, too, estimates made for a breeding group that has been subjected to a wide range of management conditions, which increases the total phenotypical variance, will have lower values than those obtained for a genetically similar group that has been maintained in a more uniform environment.

In other words, a heritability estimate is specifically a value for the breeding group from which the data were obtained, and one is not entitled to assume that because a given value has been obtained for growth rate in Charolais cattle that the same value will be obtained for this trait in another breed or even another group of Charolais. Nevertheless it is possible to say in general terms that certain characteristics are highly heritable, that some are only moderately heritable, while others are weakly inherited. As pointed out above, reproductive traits are weakly inherited but milk yield, and with it weaning weights, has a moderate value, while growth rate has a high heritability.

A knowledge of heritability values for the economic traits is important for making selection decisions. Obviously a pedigree breeder would be limiting his scope for selection in other more important directions if he tried to improve fertility by selection because additive genetic variance for this trait is low. If he has a herd fertility problem that cannot be corrected through veterinary or nutritional measures, then he is advised to introduce an outcross in the expectation that there will be some manifestation of hybrid vigour. On the other hand, where heritabilities values are high and the characteristics are economically important, breeders can reasonably expect to make worthwhile progress through selection on the basis of performance.

PERFORMANCE MEASUREMENT

There has been a profound change in selection procedures in livestock breeding over the past 50 years and it has arisen through a greater understanding of the nature of variation in important economic characters and how this variation can be exploited to best advantage in making selection decisions. A key factor in this change has been the growth of recording and it is with those animals, where recording can be reasonably precise and an easily accomplished undertaking, that the greatest progress has been made, for instance laying hens and dairy cattle. Beef cattle and sheep have been at the other end of the recording scale and breeders with these two kinds of livestock have continued to place great reliance on subjective appraisal of breeding animals but dairy cattle breeders, while continuing to pay attention to important aspects of conformation, have placed an increasing emphasis on milk yields, quality and ease of milking.

As records accumulated and analyses were made of the relative performance of close relatives, for instance daughters and dams and sib and half-sib groups, it soon became clear that the confidence that was being placed on individual records was not always justified. The emphasis in selection gradually moved away from individual to collective performance which in the first instance stressed the importance of progeny testing and the extended use of sires that had proven their worth as breeders. Ironically this, after a century and a half of emphasis on individual merit, was a return to the philosophy and practice of Bakewell, the master breeder who established the Dishley Society in order to promote progeny testing as a tool for the improvement of his Leicester sheep.

Progeny testing was also the principal tool of the Danish pig breeders in changing their native Landrace from a lard-type pig to its present form as a bacon pig. The Danish success inspired the development of a similar system in Britain with the establishment in the 1950s of 5 pig progeny testing stations to service the country's pedigree breeders. They made some progress towards desired objectives but they had several important drawbacks, notably they were expensive to run, relatively few sires could be tested in any given period because of the systems demands on accommodation and facilities, and the time taken to complete the test. Fortunately it was possible to determine from an analysis of accumulated data, not only in Britain but also in the USA that the most important characters, namely rate of growth, economy of food use and eye muscle area (established by ultrasonic measurements) had sufficiently high heritability estimates to justify the replacement of progeny testing by performance testing. The loss of accuracy through abandoning the more precise progeny testing and adopting performance testing in its stead was more than compensated for by the larger numbers that could be tested and with this, a far greater intensity of selection was possible. Moreover, performance testing reduced the generation interval and this has an important effect on rate of genetic gain.

As pig improvement policies were evolving there were corresponding changes in dairy cattle improvement but here the emphasis remained on progeny testing. There are good reasons for this. First milk yield does not have a high heritability rating and second the expression of this trait is limited to females, so that the only reasonably accurate method of assessing a bull's transmitting power in respect of milk yield is by measuring the yields of his daughters. Yields of a bull's half sisters are a useful index of his prospective worth, but the real test comes when his daughters come into production. Inevitably this means that a bull is at least 5 years old before there is even a partial assessment of his breeding value for milk yield. Though this takes time it is fortunately not an expensive operation for it is part and parcel of dairy farming that heifers enter herds each year as replacements. If these heifers are identified by sire and their yields are recorded at a little extra cost a picture of the bull's worth gradually emerges.

The dairy industry in Britain, as in many countries abroad, is fortunate that milk recording and artificial insemination, an allied tool in herd improvement, are in the hands of a national body with overall responsibilities in the field of dairy herd improvement. The English Milk Marketing Board has progressively, since 1954, built up a very efficient system of selecting promising young bulls, recording their progeny in a number of herds, and comparing their performance with contemporaries in the same herds, not only in respect of milk yield and composition but also of other important aspects such as temperament, ease of milking and teat and udder shape. Once superior sires have been recognised their value is spread widely by artificial insemination not only in commercial herds but also, with really outstanding sires, in pedigree herds.

Breeders' attitudes in both pig and dairy production changed radically as they realised the value of the above developments, but change was very slow to come on the beef side and even now attitudes in beef cattle breeding lag behind those in other production fields. There are several reasons for this. One is that there was no recording of beef animals of any great consequence, apart from the efforts of the Lincoln Red Society, until 1964 when the Beef Recording Association started operations. Another is that the typical pedigree beef herd in Britain is relatively small and each individual is easily recognisable and is known for its performance. Under these conditions there is less incentive for recording than there is with large herds. It is significant in this connection that the Lincoln Red herd is characterised by large herds. A third factor, particularly with the more fashionable breeds was the emphasis on so-called 'breed type' and its measurement, by its very nature, has to be subjective. Excellence in this respect was rewarded by show ring awards and high prices at the leading breed sales, such as those at Perth for the Aberdeen Angus, that also catered for an overseas clientele. Breeders who achieved high standing in these areas came to be regarded as master-breeders and naturally they were not interested in such innovations as selection based on performance recording. Moreover many of these successful breeders were on the ruling bodies of breed societies and their complacency extended to the policies of these societies.

PERFORMANCE TESTING IN BRITAIN

When the Beef Recording Society initiated performance testing as a service for pedigree breeders in 1964, with an intake of 36 bulls to the Yorkshire Agricultural Society's permanent showground at Harrogate, it had to do a lot of tactful pleading and cajoling to get the Society's support. It eventually came but the bulls generally did not come from the top breeders who were inclined to sit on the fence. Fortunately the first test was notably successful. Not only did it provide experiences of the problems of performance testing but it gave breeders an insight into its value and also confidence that their bulls would be properly treated at a very critical stage in their rearing. This latter point is important for no breeder of standing would wish to put any of his young bulls at risk. Most importantly, the first test effectively made the point with breeders that performance testing was not intended as a substitute for their skills, but as a source of additional information based on contemporary comparisons, on which they could make selection decisions.

This first test was primarily concerned with weight for age which American work, mainly covering British breeds, had shown to be a highly heritable character. Preston and Willis (*loc. cit*) report the very high figure of 0·70 for final weights at the conclusion of performance testing. The BRA chose as its criterion weight at 400 days rather than daily liveweight gain during test, because the bulls then coming under test at 7–8 months of age all had different rearing histories with possibilities of compensatory growth complicating the picture. A period of 6 months on a standard diet is a great leveller and the 400-day weight then becomes a reasonably reliable index of growth performance.

The first Harrogate test had a profound effect on attitudes in the Hereford Breed Society and it made substantial contributions from its own resources to the setting-up of a performance testing station for Hereford bulls at Holm Lacy. This is now administered by the MLC which has four other centres at Ingliston, near

Edinburgh; Stockton on Forest; Stoneleigh; and West Buckton, near Taunton. The annual intake varies between 400 and 500 bulls which are tested for 7 months on a standard pelleted feed supplemented with an allowance of a small quantity of long hay to promote rumination. The principal measurements apart from weight are feed consumption and backfat thickness (determined by ultrasonic meter at 12 months) and height at the withers. In addition, on completion of tests, bulls are assessed for conformation using a type-classification system.

Beef breeding in Britain, as in North America and other countries of the New World, has come a long way since the 1950s. So far as Britain is concerned there was no great progress until there was a national body, first the BRA and then the MLC, to organise and manage the necessary services and to function as the MMB and the Pig Industry Development Authority had done for their respective industries. It is interesting to note, however, that most support for performance testing comes from the owners of imported breeds. In 1982–83, for instance, just over 70 per cent of the 400 bulls that completed tests were Charolais, Simmental and Limousin representatives. The Hereford, with 12 per cent of the total, was the only British breed of major consequence in that year.

PROGENY TESTING

MLC services for beef breeding have not been limited to performance testing. A bull proving scheme is now in operation in collaboration with several breed societies, in an endeavour to spread the influence of outstanding young sires (judged by performance testing) by making them available to pedigree breeders through AI. In 1983 there were 24 bulls from 5 breed societies participating in the scheme and some representative bulls are listed in Table 4.1 with details of their performances.

Not all the bulls in the scheme are the products of central performance testing. The MLC has also organised a programme for on-farm testing for herds that can enter a minimum of 10 contemporary bulls. All the measurements made at central performance testing stations apply to on-farm testing except food consumption.

In 1983 over 450 bulls were perfomance tested under farm conditions and 5 of the 24 young bulls in the proving scheme in that year came through the on-farm screen. Obviously its application is limited to large herds and the MLC now records about 90 herds annually that have more than 40 cows. About a third of this number are Hereford herds.

The overall aim of the bull proving scheme is a co-operative

TABLE 4.1. REPRESENTATIVE BULLS IN MLC PROVING SCHEME, 1983

Breed and bull	400-day weight	kg feed/kg gain	Backfat (mm)
Charolais			
Cranebrook Stewart	714 (+63)	5·4 (−0·8)	2·2 (0·0)
Bruno Sadat	733 (+103)	7·1 (−0·3)	1·6 (−0·2)
Hereford			
Grimshaw I Slam	557 (+60)	5·5 (−1·5)	4·7 (0·0)
Kingsland 1 Tintagel	546 (+74)	na*	3·1 (0·4)
Limousin			
Newclose Solomon	573 (+45)	5·3 (−0·6)	1·8 (−0·5)
Wilkesly Samson	588 (+69)	5·4 (−1·1)	3·2 (0·2)
Simmental			
Ganley Lysander	703 (+73)	7·0 (−0·5)	3·2 (+0·5)
Hockenhall Magnum	740 (+126)	6·9 (+0·1)	3·5 (+1·0)
Welsh Black			
Haulfryn Byron 5	528 (+55)	6·3 (+0·1)	2·6 (−0·6)
Tyddewi William	622 (+75)	6·5 (+0·4)	2·3 (−0·5)
	(+heavier)	(− less feed)	(− leaner)

(* na = not available)

effort with breed societies to spread the influence of good bulls by making them available through AI to pedigree breeders. The scheme is still in its early stages but the testing procedure aims to produce 35–40 calves by each bull in a number of herds, along with contemporary calves by other bulls to provide comparative growth data. Essentially this is an application of a technique of an on-farm comparison, which has been notably successful in dairy herd improvement in producing a steady supply of superior proven sires which can be exploited by artificial insemination. This incidentally has not been greatly favoured by some of the beef breed societies. Indeed there was a time in some breeds where animals bred by AI were not accepted for registration as pedigrees. This short-sighted view, prompted by fears of leading breeders that AI would significantly reduce the demand for their bulls, is now generally recognised as being detrimental to a breed society's efforts to compete with other breeds. Artificial insemination, as the dairy industry can bear witness, is a most potent tool in breed improvement if it is sensibly applied.

Interestingly, it was the MMB and not the MLC that pioneered progeny testing of beef bulls. The Board is a very interested party in beef breeding because of its dominant role in the organisation of artificial insemination. In England and Wales alone, there are approximately three-quarters of a million inseminations, or approximately one-third of the total, that are accounted for by beef breeds. Most of these relate to dairy cows but there is a sizeable proportion of beef cows also included in the total. AI was the method by which the Charolais was first used in Britain when it was being tested and AI continues to be very useful for breeders, particularly those concerned with imported breeds, who have so few cows that natural service is uneconomic.

The MMB scheme, which is based at its Warren Farm, has been in operation since 1970 and covers the principal beef breeds, but the greatest emphasis has been on the Hereford which accounts for about half of the beef inseminations. The usual procedure, following those for proving dairy sires, is to have teams of young bulls that have been selected on available criteria such as performance test records. Sufficient matings are made by artificial insemination for each sire on test to provide representative groups of male calves that are assembled at Warren Farm at approximately 10 days of age. Each intake extends over a 3-week-calving period so that sires are represented by compact groups with the same general environmental conditions. Eventually the steers are finished on a grass-conserved fodder regime which is the method most widely followed with dairy-bred beef.

The testing programme covers the following aspects: (a) dystocia and calf mortality, (b) liveweight gain to slaughter, (c) carcass weight and dead weight increase of progeny groups and (d) carcass evaluation. This last aspect is handled on behalf of the Board by a unit in the MLC which specialises in carcass analysis and appraisal. Problems associated with parturition are of particular consequence in the larger imported breeds, notably the Charolais and the Simmental as reference to Table 3.5 will confirm. The Board has paid particular attention to this point starting with the trial importation of Charolais bulls. The earliest assessment showed a very considerable variation between bulls in respect of difficult calving and some which had particularly bad records were destined for immediate slaughter. The monitoring of sires for calving difficulties by the MMB and by other organisations with a responsibility for artificial insemination makes an important contribution to the usefulness of these new breeds. Commercial farmers, both in dairy and suckled-calf production, cannot afford excessive wear and tear on their cows or loss of calves at birth and ease of calving must be emphasised in selection programmes. This might be at the expense of weight for age but this would be a price worth paying for improvement in this respect.

RATE OF GENETIC GAIN

There is no point in assembling herd or breed records unless they are used to good effect in purposeful breeding programmes. There are signs that this is happening on an increasing scale in beef cattle breeding and this reflects a more objective approach to selection, especially of bulls as a result of a growing appreciation of basic principles of population genetics. The old concept of breeding based on phenotypic excellence no longer has much support, for the worth of a parent cannot be measured by looks alone. The real test is the performance of offspring and in this it is the male parent that has the major role because of the much greater influence he has on progress in a herd as compared to the individual cows to which he is mated.

Genetic progress for any given character depends on three main factors:

(a) The selection differential (SD) which is defined as difference in performance of a selected parent over the average performance of the breeding group to which it belongs.

(b) The accuracy of the selection in genetic terms. A differential is only meaningful if it has been established under the same environmental conditions because one that is mainly attributable to superior nutrition, for instance, has little validity in selection. Here heritability values are important because they are indicative of the genetic value of observed differences since they are a measure of the additive genetic variance which the breeder aims to exploit.

(c) The generation interval. This can be defined as the time that the average breeding animal remains in a herd or flock. In the case of bulls this may be relatively short, perhaps no more than 2 years, but beef cows generally have much longer herd lives and averages of 7–8 years are by no means uncommon.

In mathematical terms, the equation for genetic gain in respect of a given measurable character is:

$$AGG = \frac{S \times h^2}{I}$$

where AGG is annual genetic gain,

        S is the selection difference,

        h^2 is the heritability of the character,

        I is the generation interval.

It can be illustrated by an example relating to the 400-day weight of a selected sire which in a performance test is 80 kg heavier than his contemporaries, who are regarded as averge for a breed where the heritability estimate for their character is 0·5. The bull is used for only 2 years when he is replaced by a second bull, so his generation interval has a value of 2.

$$AGG = \frac{80 \times 0\cdot5}{2} = 20 \text{ kg.}$$

It must be remembered though, that the bull is only responsible for half the genotype of his offspring so if he is mated to average females the gain is 10 kg and not 20 kg per annum. This figure, of course, applies only to the bull. Some gains are possible on the female side as well, but these will be appreciably less than those from the male. This is due to (a) the lower scope for selection where at least half the heifers born in a herd are required for herd replacement, (b) the longer generation interval for females and (c) the stress that has to be placed on characters other than weight-for-age. In practice a breeder would be fortunate in securing greater gains than 1–2 kg annually (varying according to sex) in 400-day weights based on a within herd performance test, where it is not possible to subject all candidates for selection to the same environment as is the case with station-tested animals.

The above examples illustrate the much greater importance of the sire in determining genetic gain, as compared with his mates, both individually and collectively. Primarily this is due to the greater intensity of selection that can be practised on the male side but there are other considerations as well. It is feasible, and it can be advantageous also to shorten the generation interval on the male side by a rapid turnover of sires, always providing the replacement sire is potentially a better prospect than the one he succeeds. This is not generally practicable on the female side because of the greater impact of wastage factors such as infertility, mastitis or a variety of husbandry considerations. Once a cow has established herself as a worthy herd member by virtue of regular and easy calving and better than average offspring there is little point in replacing her until it is absolutely necessary with what, at best, is still an untried heifer.

The argument can follow similar lines with a bull, namely why replace him while he is still fertile and producing good calves? But the aim must always be one of replacing a good sire with an even better one. This is where the breeder with a really large herd is at an advantage for he has scope for testing potential stock sires without committing the herd too deeply. Where herds are small, as they are for the most part in Britain, the case for co-operative breeding is unassailable and this is where developments in the dairy industry have a particular significance for beef breeders. Apart from the Milk Marketing Board's work in testing promising young bulls and then selecting the best for extended use by artificial insemination, there are also influential groups of Friesian breeders who have combined in a programme of sire testing and then making semen from top bulls available to all co-operating breeders. Even the owners of large herds have come to realise that they can make only limited progress working on their own.

Breed improvement in farm livestock is essentially a numbers game, just as it is in plant breeding. Apart from the need for large breeding populations in order to apply the principles of population genetics to best advantage, testing is an expensive operation; whether it is on the basis of individual performance for characters that are strongly inherited and can be measured in the living animal or by progeny performance for characters of low heritability, characters expressed by only one sex or characters that can only be determined satisfactorily after slaughter. This is particularly true of progeny testing and it makes sense in any breeding programme to use performance testing as the first screen to determine the best prospects for the more expensive and time-consuming progeny testing. Once the really outstanding sires have been recognised by virtue of their breeding performance then they should be used to the greatest possible degree and this, as has been stressed previously, is where artificial insemination comes into its own. Selection must, however, be a continuing process and it is not enough to create an elite group of sires. There must be a following generation in the pipeline of even greater genetic merit to take their place if genetic progress is to be sustained in any breed.

Chapter 5

BREEDING AND SELECTION (2)

Selection Objectives

It is important that pedigree breeders are realistic in their selection objectives and that they have a sense of priorities because every additional consideration reduces the intensity of selection in other directions. There have been examples, fortunately now belonging to the past, where a pre-occupation with trivial points did positive harm to breeds; for instance the six white points of Berkshire pigs and the 'vessel' type of the Ayrshire, where a flat-soled udder became such an obsession among some breeders that two schools grew up within the breed.

There are still reminders of this past which can be termed 'the Crufts' approach' to cattle breeding. In 1969 at the official opening of the Aberdeen Angus performance testing station near Aberdeen, which coincided with the termination of test for the first intake of bulls, a leading breeder obviously referring to an extremely well-grown bull that had excited a lot of comment and admiration, spoke of the danger of losing the 'true' Aberdeen Angus type; but what is this type? Is it the sort of bull people hope to breed in order to hit the jackpot at the Perth sales because it appeals to other breeders who have similar ideas and the money to back their confidence in them; or is it the type of Angus that the commercial man wants to improve the profitability of his beef enterprise? Surely it must be the latter and, if this is so, the ideal type will evolve when maximal progress has been made in those economically important traits!

The attainment of these goals involves measurement and recording so that wherever possible there is objectivity rather than subjectivity in making selection decisions. The same man on this occasion referred rather scathingly to 'paper breeders' and implied that there was some divine gift of the privileged few that made them the master breeders. One does not deny the importance of an eye for stock and a sensible balance in selection, because the fastest growing bull in the world will be of little value if he has some serious congenital shortcoming that can be passed on to his progeny, but equally, the prettiest bull that ever graced the Royal Highland has no future if his offspring are mean performers where it really matters.

Fortunately the old conflict between geneticist and breeder is now sporadic as each party gets a better understanding of the other's viewpoint. The BRA, for instance, always took the attitude that weight for age was just one consideration. All it set out to do, when it established performance testing stations and started on-farm recording, was to provide additional information for breeders. It did not suggest that Aberdeen Angus or Hereford breeders should enter into a race with the Charolais, which already had a big head start, to produce fast-growing, late-maturing cattle that are well suited to finishing on high energy diets. Decisions as to goals belong to breeders, and in the case of the two breeds mentioned above they could be well advised to concentrate on early-maturing, but nevertheless fast-growing cattle to supply what is bound to be a continuing need, namely stock that will finish on pasture or high roughage diets.

BREEDING AIMS

The list of possible breeding aims in beef cattle is formidable and here are some of them, not in order of importance because this will vary from one breed to another according to function:

(a) Growth rate.
(b) Efficiency of food use.
(c) Adaptation to particular environments.
(d) Rate of maturity.
(e) Fertility and regularity and ease of calving.
(f) Longevity in breeding stock.
(g) Carcass attributes.
(h) Milking ability.
(i) Killing-out percentage.
(j) Breed type.

As was pointed out in the previous chapter some of these qualities will show little response to selection, for example longevity, fertility and regularity of calving (because of low heritabilities) but in a sense they look after themselves because a shy breeder or a short-lived cow does not leave many offspring behind. Others such as growth rate are not only highly heritable but are also measurable.

A number of the features are not easily definable, for instance adaptation. Where this refers to suitability for harsh environments obviously there is a complex of contributory traits and all that a pedigree breeder can do is to select animals under management conditions appropriate to their commercial function. The Cadzow family have demonstrated this effectively in their handling of the Luing, but one comes across cases where pedigree Galloway stock are kept under conditions which are well away from their normal habitat. However, the Galloway Society, very wisely, has not participated in the national performance test programme in its present form because it does not suit the requirements of the breed. Fortunately there are a number of large Galloway herds, and on-farm recording is prospectively very valuable with them because there are the numbers to permit meaningful contemporary comparisons being made in an environment that is normal to the breed.

Milk yield is very important in those breeds which are used as pure breds or in crosses with other beef breeds for suckled calf production. Complaints have been made in Australia that the traditional British beef breeds have insufficient milk to rear good calves. Here in Britain such a deficiency in the more popular beef breeds is not so serious because of the greater use that is made of beef × dairy cows in suckler production. Milking ability must, however, continue to be an important criterion in the Galloway, Welsh Black, Devon, Sussex and Lincoln Red that are used as purebreds for suckler production and the most suitable measurement of this attribute is the adjusted 200-day weight of the calves.

CARCASS AND MEAT QUALITY

It is almost impossible to define carcass quality so that it can be assessed reliably in the living animal. The classic beef type is a short-legged, blocky animal, with a high spring of rib and well-developed loins and hindquarters. Emphasis is placed on a full back-thigh and an even distribution of subcutaneous fat in the finished animal without any hint of patchiness. The inference of a high spring of rib, which gives a round cross-section as opposed to the wedge cross-section characteristic of the extreme dairy type, is that it permits a full development of the eye muscle (*longissimus dorsi*), which is the main muscle of a rib roast and one of the most valuable parts of the carcass.

Work undertaken by Butterfield* in Australia covering a range of breeds and crosses, and also animals within breeds that have been subjected to contrasting rearing regimes, indicates that there are surprisingly small differences in the balance of expensive

---

* *Carcase Composition and Appraisal of Meat Animals* (ed. D. E. Tribe) C.S.I.R.O., Melbourne, 1964.

muscles between widely differing types of cattle. There are, of course, very big differences in fat composition which are attributable to both genetic and nutritional factors and much of the believed superior conformation of the recognised beef breeds, when they are judged on the hoof, is due to fat cover. This point will be discussed in greater detail in Chapter 7 dealing with growth and development.

There are differences between breeds in the way that fat is laid down. The improved beef breeds at point of finish for slaughter have more intramuscular fat and less channel fat than unimproved breeds like the Friesian, and this will affect the cut-out values of the resulting carcasses. Though there is no really accurate method of determining differences in the distribution of fat and musculature in the living animal, nevertheless it is possible to get some estimate of fatness by ultrasonic measurements at selected sites along the back line. The technique is now reasonably reliable and as indicated in Chapter 4 backfat measurements are now an integral part of the MLC performance testing programme. Ultrasonic scanning is also useful in selecting animals for slaughter in experimental studies where the aim is slaughter at a standard finish.

THE BUTCHER'S NEEDS

Indeed in the present state of knowledge and of flux in the butcher's requirements to satisfy his own pocket and the needs of his customers, it is not easy to define selection ideals in respect of carcass attributes. Work at Newcastle University on consumers' preferences indicates that tenderness is the one attribute that is consistently recognised and appreciated but this is a function of age, cut and conditioning rather than heredity. There is also a very marked preference for lean meat.

The butcher's prime consideration after satisfying his customers' requirements is the highest possible cut-out value in the carcasses he buys. This apparently is much less affected by breed *per se* than by the finish of the carcass. Overfat carcasses which necessitate a lot of trimming will obviously be less valuable to the butcher than those that require little or no trimming; but at the other end of the scale unfinished carcasses which have a high percentage of bone relative to edible meat (muscle + saleable fat) will also have low cut-out values. These, however, are considerations that are more under the feeder's rather than the breeder's control.

With the growth in demand for grilling, frying and stewing cuts of beef as opposed to roasts, about the nearest one can get to defining breeding objectives for carcass attributes is a maximal development of muscle with as high a proportion of this muscle being located in those parts of the carcass that command the higher prices. This in effect means a relatively late-maturing animal, as opposed to the early-maturing types that were favoured by breeders up till a few years ago. Degrees of fattening, to suit differences that exist in the trade, can then be met by feeders who are able to control both plane of nutrition and point of slaughter. There will be no conflict between these objectives of more lean meat and two other primary selection objectives, namely rate of gain and economy of food use.

FUNCTIONS OF BREED SOCIETIES

The role of breed societies is now very different from what it was 25 years ago when their principal activities were the maintenance of herd books, breed promotion and safeguarding the interests of members. In this last area policies were often reactionary and detrimental to the advancement of a breed and there is no better example of this than the limitations some societies placed on the application of artificial insemination. Long after dairymen had accepted AI as an important means of betterment in their respective breeds some beef societies were still placing severe restrictions on its use. At one time the Hereford Society refused to accept any animals for registration that were bred by AI with one exception, namely where a bull owned by a breeder was incapable of natural service. Later jointly-owned bulls became eligible and as a major concession, what were termed 'proven high performance progeny bulls' standing at AI stations achieved respectability but still with limitations on their use. Only 2 bulls were nominated for use at any one time and there was a limit of 6 registrations per herd annually. On top of this there were what were then swingeing registration fees with bulls costing £25 and heifers costing £15 to provide further discouragement of artificial breeding.

This attitude was not only selfish but short-sighted so far as the long-term future of a breed was concerned, especially in a situation where there is a high proportion of small herds to limit the in-herd application of modern concepts of breed improvement. It stemmed from the control of a society's policy being largely in the hands of top breeders whose bulls had been commanding high prices for pedigree breeding purposes. It is understandable that such people would wish to preserve the *status quo* which was to their individual benefit though it was not to their breed as a whole.

Fortunately there has been a great liberalising of attitudes and controls. This was accelerated by the introduction of Continental breeds and with this a growing realisation that the more important form of competition was not between members of a breed society but between breeds. The dairy industry had seen the almost total eclipse of the once dominant Dairy Shorthorn by the British Friesian and the lesson, plain for everyone to see, was that popular beef breeds could go the same way. Beef Shorthorns were the first to feel the impact of competition and here the principal reason was the preoccupation with the compact early-maturing type once greatly favoured abroad, at the expense of growth rate and substance. Breeders were too late in seeing the writing on the wall and now the Beef Shorthorn qualifies for the rare breed category. It is symptomatic of its present day status, that invariably there are now more Longhorn than Beef Shorthorn entries at the Royal Show and one has to go to the Great Highland Show to see a sizeable Beef Shorthorn representation. Incidentally the Aberdeen Angus at either show or Smithfield is only a shadow of what it was 20 years ago. The really impressive displays at major shows are now provided by the Continentals, in particular the Charolais and the Simmental with the Limousin also making a substantial impact.

The societies representing these three breeds have been particularly active in the MMB-controlled beef progeny testing programme at Warren Farm and they are also strongly involved in a Young Bull Proving Scheme which is administered by the MLC. The Limousin Society has gone a little further than the other two and has devised what it calls the Lim-Elite Scheme where young superior performance-tested bulls, included in the MLC Young Bull Proving Scheme covering purebred matings, are given additional matings in Friesan herds to provide information on ease of calving as well as growth and carcass attributes in their crossbred progeny. The ability to store deep frozen semen adds greatly to the value of such testing, for once really outstanding sires have been recognised they can be widely exploited by building up semen banks for subsequent use.

It is not only the Continental breed societies that have made good use of the services offered by the MLC and the MMB. Herefords were represented right from the outset in the Warren Farm programme but here the first objective was less breed improvement than the recognition of outstanding sires for cross breeding in dairy herds. Over two-thirds of the beef inseminations now come from progeny-tested sires and Hereford semen accounts for approximately half of these, totally over 200,000 inseminations annually. The Hereford × Friesian is one of Britain's most important feeding animals and the continued MMB emphasis on the Hereford is fully justified. The value of the MMB programme is now more widely realised by the Hereford Society which also supports the MLC Young Bull Proving Scheme.

The first British breed society, however, to develop a co-operative breeding plan based on performance testing was the South Devon Society. Because it was the largest of the British breeds it excited interest abroad, particularly in North America which was in the forefront of the move away from early-maturing breeds. As a result the top bulls in the South Devon performance testing programme were in great demand and they were being exported before they were able to leave any progeny behind them. This was an expensive loss, akin to consuming the seed corn, and though the individual breeders concerned profited, their good fortune was at the expense of the breed as a whole and also of the MLC. Breeders' contributions did not cover the costs of performance testing and the MLC had to make good the difference, but fortunately a solution was reached with the support of the Breed Society. It was agreed that performance-tested bulls with good records should be subjected to semen tests and where semen was of good quality sufficient should be collected and stored to provide about 400 straws for subsequent use in pedigree herds. The overall aim has been one of establishing the relative merit of a bull's progeny and the entry of bulls resulting from these matings in further performance tests. Altogether the concept is an excellent one and an example for other breeds to follow. Probably it would have received much more prominence had it not been for the mushroom growth in popularity of the Charolais and Simmental with marginally better weight for age performances than the South Devon.

CROSS BREEDING

A high proportion of animals that are deliberately bred for beef production in Britain are cross-breds and there are good reasons for this. First, there are the advantages to be gained from compensatory mating, which are illustrated by the use of beef bulls on dairy cows in order to improve carcass attributes. At one extreme the pure Jersey has little to recommend it as a beef animal and therefore nobody in his right mind is likely to rear a Jersey bull calf for beef purposes. A Charolais × Jersey, on the other hand, is quite a respectable beef animal and thoroughly justifies being used for this purpose. Most matings with dairy cows are less extreme than the Charolais × Jersey; the Hereford × Friesian combination is particularly popular both with dairymen for the premium these calves earn over the pure Friesian, and also with the feeder because of its performance on pasture and high roughage diets as well as its better grading prospects.

Another point of considerable importance, especially with first calving heifers, is relative ease of calving. There are considerable risks of dystocia in Friesian first calvers that have been steamed up to give a good start to lactation and for this reason many dairy farmers prefer not to mate their maiden heifers back to the Friesian but to use the Aberdeen Angus, which among British breeds is at the top of the league for ease of calving and neo-natal deaths (see Table 3.4). It is followed by the Hereford which also has yet another virtue in that it is a very docile breed and this has much to recommend it where a bull is running with a bunch of heifers.

A third reason for cross breeding is the exploitation of hybrid vigour, a phenomenon which can be described as a performance in cross-breds that is better than the average of the two parent breeds. This can be illustrated with 2 breeds of hill sheep which both have a lambing performance of 120 per cent under the same management but when mated together their cross-bred offspring, again under the same management, average 130 per cent. The extra 10 per cent in effective lambing percentage over expectation is attributable to hybrid vigour. Hybrid vigour is of the greatest importance in the so-called 'fitness traits', such as reproductive performance and survival power. It has less significance in strongly inherited characters, such as growth rate or efficiency of food use which usually conform to the average of parent breed values. The greater ability of cross-breds to survive and thrive, as compared with purebreds, has long been recognised by stockmen and it has been used to great advantage in the pig industry where the cross-bred or hybrid sow is well recognised for its superiority over the purebred in respect of litter size and piglet survival.

Another area where the value of cross-breeding has even longer recognition is in sheep breeding where the combined effects of compensatory mating and hybrid vigour are exploited in a stratified system of land use in Britain. Upland breeds such as the Swaledale and Scotch Blackface are well adapted to harsh environments, though with comparative low levels of productivity. Draft ewes, when moved to easier country and mated to long wool rams such as the Border Leicester, produce useful cross-bred lambs with castrate males going for slaughter, usually off forage crops. The females, however, are much too valuable for immediate slaughter because they are first class mothers well suited for intensive fat lamb production. Unquestionably the merit of these cross-bred ewes owes a lot to hybrid vigour. It is not exceptional for an age balanced flock of Mule or Greyface ewes under good management to rear better than 180 lambs for every 100 ewes mated. Their mates are generally a third breed, such as the Suffolk or the Hampshire, which is chosen because of its carcass attributes so that the resulting slaughter lambs are the product of a three-way cross.

There is a similar pattern of land use in beef production. The Galloway cow, like the Blackface, is well suited to difficult upland conditions but this adaptation is at the expense of growth and rate of maturity in its progeny. Though pure Galloway steers eventually make attractive carcasses, with the breed boasting a Grand Championship at Smithfield a few years ago, they are, on the whole, very slow feeders. Consequently the aim with many commercial Galloway breeders is to limit pure matings to the minimum required for herd replacements. The favourite crossing bull is a white Shorthorn and though the resulting steers are good feeding animals the more important product is the Blue-Grey female which is widely regarded as the best suckler cow under marginal farming conditions and as one might expect it is particularly popular in the Borders and further north in Scotland. Important among its virtues are hardiness (a legacy from the Galloway), regularity of breeding and longevity. It is not uncommon for a Blue-Grey to produce more than a dozen good calves in its working life-time and the MLC surveys show that an average Blue-Grey herd life of slightly under 10 years is higher than that for any other cross-bred cow.

The most popular bull for use on the Blue-Grey used to be the Aberdeen Angus and though the resulting beef was a very high quality there was not enough of it to please the feeder once greater emphasis was placed on weight-for-age. From about 1960 there was a move to the Hereford and though the calf crop varied in colour from red to black with degrees of roaning in between, there was always the characteristic Hereford marking and with it some assurance that they would grow faster and make greater weights than Aberdeen Angus cross-breds. From the early 1970s Continental breeds, in particular the Charolais, have largely taken over from the Hereford in the three-way cross of the slaughter generation.

Still on the female side another highly regarded cross is the Hereford-Friesian which now is possibly the most important suckler in numerical terms. Unlike the Blue-Grey, where the heifer is the primary reason for crossing, the Hereford × Friesian female, until its merit as a suckler was recognised, was and still is to some extent regarded as an incidental by-product of breeding bull calves which are worth about 25 per cent more than heifer calves at the week-old stage. Some suckler producers criticise the Hereford × Friesian cow on the gounds that it can produce more milk than a young calf can cope with and compare it unfavourably in this respect with the

Dairy Shorthorn-cross cows, either by the Hereford or the Angus that used to come in considerable numbers from Ireland. Be that as it may there is no question but that the Hereford × Friesian cow, when mated to one of the larger Continental breeds produces truly magnificent weaners. It is not exceptional for these at only 9 months of age to have a liveweight of 360 kg and sometimes more. They provide a first class example of how a judicious use of cross-breeding can, in two steps, transform excellence in dairying to excellence in beef production.

The earlier popularity of both the Hereford and the Aberdeen Angus for top crossing on dairy cows unquestionably owed a lot to their ability to colour mark. It is not difficult to recognise a cross Hereford because of the dominance of the Hereford pattern while the dominant poll and black factors in the Aberdeen Angus together put an unmistakable label on its cross-breds. This was particularly valuable to farmers when the Government was paying rearing subsidies on nine-month-old animals deemed suitable for beef. Beyond this, colour markings provide purchasers with a measure of confidence in the potential of animals on offer to them. Fortunately for this breed the Charolais also puts an unmistakable label on its cross-breds and this is also the case with the majority of Simmental crosses. Unfortunately for the Limousin it does not leave such a positive mark and here the buyer of a week-old calf has little more than the word of the vendor to establish that the calf on offer has a Limousin sire. It is different in the fat stock ring because the Limousin attributes are then most easily recognised by the discerning buyer.

IMPROVEMENT OF BEEF QUALITIES IN DAIRY BREEDS

Because a substantial proportion of the clean beef produced in Britain comes from pure and crossbred animals that originate in the dairy industry, it is sensible to establish whether beef attributes in dairy breeds can be improved without detriment to dairy qualities. About 1956 V. E. Vial, of King's College, Newcastle, and I. L. Mason, of the Department of Animal Genetics, University of Edinburgh, co-operated in a study of beef attributes of male progeny of Friesian sires standing at AI stations to find out whether there was any conflict between the beef performance of these animals and the dairy performance of the female half sibs which was established by the normal method of contemporary comparison.

The investigation was not as complete as one would have desired because there were insufficient progeny in the groups and insufficient money in the kitty to permit a fuller assessment of carcase attributes than it was possible to make under the circumstances. However, there was no evidence of any serious conflict between the two aims, and fast-growing Friesians, it seems, are just as likely to be good or bad milk producers as slow-growing ones. Preston and Willis in their review of the literature on this topic state 'The balance of the evidence indicates that the relationship between meat and milk is either zero or slightly positive. In practical terms one can conclude that selection for either of these traits will not be detrimental to the other and may even improve it slightly.'*

A Beef Improvement Study Group, organised by the MLC, suggested in 1971 that the Milk Marketing Board should embark on a programme of performance testing for 600 bull calves from their nominated matings scheme, and then selecting the top third for growth rate and efficiency of food utilisation as the annual intake for their dairy progeny testing scheme which determines which bulls are worthy of entry into the Board's top AI stud. It was calculated that there could be a 0·35 per cent annual increase in 400-day weights arising from this programme. In absolute terms this would be about 1·6 kg of gain annually for a bull calf and though this is not a very large increase, nevertheless it amounts to a lot of beef with an annual output of nearly three-quarters of a million Friesian beef calves.

The dairy beef programme has not, however, developed according to a blueprint of this nature. Instead there has been a swing to the angular North American Holstein away from the fleshy Dutch type which is the British Friesian. The reason lies in the Holstein's higher milk yields and up until 1984 expansion of milk production has been the principal objective in dairying and this has been at the expense of reduced acceptability of carcasses from Holstein matings. The swing to Holstein has not been confined to the United Kingdom but applies to practically all countries in the EEC, and unquestionably the magnitude of current milk surpluses owes a lot to this infusion of Holstein blood.

It is an irony that this neglect of beef attributes has created a situation where dairy farmers asked to curtail milk production may be turning towards beef to take up some of the slack. It may well be that we will see a much greater emphasis on dual-purpose function of dairy cows than there has been since the pre-war years of Dairy Shorthorn dominance.

---

*Preston, T. R. and Willis, M. B. (1970), *Intensive Beef Production*, Pergamon Press, Oxford.

Chapter 6

REPRODUCTIVE EFFICIENCY

A FARMER, whether he is in dairying or suckler production, obviously aims for a hundred per cent calving record, but in practice it is very seldom that this figure is attained. There is almost invariably a small proportion of cows that fail to get in calf, also abortions and neo-natal deaths in calves. The Meat and Livestock Commission has collected figures of effective calving rates in suckler herds over a range of conditions and these are summarised in Table 6.1.

TABLE 6.1. CALVING PERFORMANCE IN SUCKLER HERDS 1974–75

	Lowland		Upland		Hill	
	Autumn	Winter–Spring	Autumn	Winter–Spring	Autumn	Winter–Spring
Percentage barren:						
Average	5·1	4·6	5·6	4·9	6·4	6·0
Top third	4·0	3·1	4·1	3·7	4·9	4·2
Percentage calf mortality:						
Average	5·2	6·1	5·7	5·4	6·6	5·6
Top third	3·1	4·1	3·1	4·7	5·0	4·9
Weaning percentage:						
Average	93	94	93	94	92	93
Top third	96	96	95	96	94	95

These mean figures shown in Table 8 tend to hide the fact that they stem from a series of herds with a very wide range of performance. Moreover, the data relate to one year and there are certain to be big variations, especially in times of severe winters. For example, the weaning figures for hill herds are in excess of 90 per cent, whereas in 1971 they were only 84·3 per cent. Clearly one would expect some advantage for lowland and upland farms over hill farms, particularly in adverse weather conditions. But even a figure of 84 per cent would be regarded as very high in some parts of the world, for instance in the dry veld regions of Africa or the tropical regions of South America where a cow may often produce a weaned calf only in alternate years.

In such areas there are great extremes of feeding as well as greater risks from disease and predators, but farming is on a very large scale with a much lower investment in breeding stock. In Britain with replacement heifers costing in excess of £500 at point of calving, one cannot afford to have many in a herd that do not wean a calf each year. If a cow loses its calf, then it is worthwhile to set on a dairy-bred calf, for apart from any profit the foster calf may show, this will prevent the cow from getting over-fat and with it the possibility that she will be more difficult to get in calf.

THE CALVING PROGRAMME

A herd-owner is not only interested in getting the maximum calving rate, but also the optimal concentration of calving at the time of the year best suited to the farm. The reasons for this are fairly obvious. First, there is a simplification of herd management, and, second, there is a more even bunch of weaners, either for sale or subsequent finishing on the farm. The normal gestation period in British breeds is about 283 days (286 for some large European breeds), and so the aim is to get cows in calf approximately 85 days after their previous calving. Cows seldom show obvious signs of being in season until about 60 days after calving, though silent heats can occur prior to this. However, because the uterus is not completely involuted until about sixty days from calving, service prior to this results in a lower than normal conception rate.

The usual interval between heat periods is 21 days with a variation of 2–3 days once the cycle is properly established and so there is relatively little scope for advancing calving once a cow has slipped behind her optimum calving time. At the best one can only hope to pick up 20–30 days in any one year. For this reason most farmers tend to calve their heifers for the first time a little in advance of the target date for the main herd so as to have some time in hand in case of accidents. Even with natural service, using a completely fertile bull, one is fortunate to have more than 80 per cent of a herd holding to first service. Another good management reason for calving heifers before the main herd is that they can be given more attention at calving and this is important because they are more likely to be subject to calving difficulties than mature cattle, especially if they calve down in high condition.

GENETIC FACTORS AFFECTING REPRODUCTIVE EFFICIENCY

For all practical purposes, selection within a breed will not improve fertility for the heritability of this trait is very low. This does not mean that fertility is not under genetic control for there are differences between breeds. The same is true of calf survival and for this reason the best prospect of improving calving and weaning performance as well as getting longevity in breeding cows is by exploiting hybrid vigour—in other words, cross-breeding. This can be done in two ways—mating a cow of one breed to a bull of another breed improves both calving percentage, possibly as a result of a reduction of foetal mortality, and survival rate of the resulting cross-bred calves. The second advantage comes from the use of cross-bred dams which may be mated back to one of the parent breeds or else to a third breed.

Though the value of cross-bred dams in respect of fitness traits is widely appreciated in many branches of livestock farming, notably sheep and pig production in this country, there is comparatively little published information on the advantages of cross-bred dams in beef production. Some of these can be attributed to the benefits deriving from compensatory mating rather than from hybrid vigour in itself but a very general concensus of opinion among producers of suckled calves is that they much prefer cross-bred to purebred cows because of greater regularity of breeding, longevity, and general mothering ability.

Fortunately, with the comparatively good calving performance we have in Britain, there is scope for a considerable measure of cross-breeding—up to 40 per cent in the average commercial herd—to produce first-cross heifers that can then be used for breeding a slaughter generation. There is very little scope for such a practice where the average calf crop is less than 70 per cent because all matings must be directed to maintaining the foundation breed, unless cross-breeding is a first step in grading up to another breed or, alternatively, a programme of rotational cross-breeding is being followed. There is very little information about the potential benefits of this last-mentioned course of action, because scientifically-planned trials of this have been in progress for a comparatively short period of time.

There is no prospect of increasing twinning rate by selection as one can, admittedly very slowly, in sheep, since the usual twinning rate in cattle is under two per cent and there is the added complication of free-martins. About 90 per cent of the females of mixed sex twins do not develop normally because they have been affected *in utero* by male hormones from their twins circulating in their blood streams. Taking all heifers that are born as twins, mixed or of the one sex only, the expectation is that only 60 per cent will be normal. The only hope that there is for improving reproductive rate in cattle beyond the normal ceiling of one calf per cow per year is by hormone manipulation but before considering

1. A forerunner of the modern Angus: a four-year-old bull of the Polled Angus breed bred by Hugh Watson of Keillor. From a painting in 'The Breeds of the Domestic Animals of the British Islands' by David Low, published in 1842.

2. Yearling heifers of the Luing breed. Developed by the Cadzow family from a Beef Shorthorn and Highland basis, this is the newest breed in Britain. It was evolved as a suckler cow for hard environmental conditions.

3. Hereford × Friesian heifers on good grazing. This is one of the better types of large framed animal suited to suckler calf production under good conditions.

this possibility it is necessary to give a brief account of the role of hormones in the development and control of reproductive functions in cattle.

HORMONES IN REPRODUCTION

Hormones are complex secretions which circulate in small quantities in the blood stream to stimulate certain organs and tissues to special activity. There are a considerable number of these hormones affecting such attributes as growth, the initiation of milk secretion and the let-down of milk. In this context we are only interested in those hormones which are directly or indirectly concerned with the development and functioning of reproductive organs. The master gland in this is the pituitary, a very small gland, weighing no more than three grams, which is located at the base of the brain. In the male the pituitary produces at least two hormones which affect the normal functioning and development of the testicles. Unless there is the stimulus of these hormones, a bull will be sterile. The testicles, in their turn, also produce a male sex hormone which not only stimulates the desire to mate, but also contributes to the development of the secondary sexual characteristics of the entire male.

In the female the anterior lobe of the pituitary body produces three hormones which influence sexual development; prolactin which is concerned in the growth of mammary tissue and the initiation of milk secretion, the follicle stimulating hormone (FSH) and the luteinizing hormone (LH). FSH is especially concerned with the development of the ovarian follicles in which the ova mature, while LH causes the follicles to rupture so that they are released to move into the reproductive tract where fertilisation will be effected, followed in normal circumstances by foetal growth. LH is also concerned in the growth of the corpus luteum or 'yellow body', which has an important role in controlling the oestrus cycle.

The ovaries produce the two hormones oestrogen and progesterone, which are to some extent antagonistic. Oestrogen is produced in comparatively large quantities just before the maturation of the follicle and the release of the ovum and among other things it stimulates the desire to mate. It has other functions too: in particular it influences the development of the secondary female characters such as the mammary gland, and the tone of the reproductive tract. Progesterone is produced with the development of the corpus luteum and it supresses 'heat', or the desire to mate, and stimulates the uterus so that it is in a suitable condition to receive the developing embryo.

If pregnancy occurs following ovulation, the corpus luteum

persists for at least 200 days so that there is no recurrence of heat during pregnancy with a normal female. If the heifer is not mated at a heat period the yellow body dies away after 16–18 days and oestrogen moves into the ascendancy with the maturation of another follicle and the heat period occurs once more about 21 days after the previous heat period. Sometimes two or more eggs will be shed at the one heat period and if these are fertilised then twins, or more occasionally triplets, will develop. Sometimes a fertilised egg will split and we get the production of identical twins which occurs about once in a thousand calvings.

We see from the foregoing very sketchy account of the role of hormones in reproduction the delicacy of the mechanisms that are involved in the production of a new generation. It is not just a question of ensuring that a fertilised egg can proceed through the phases of normal foetal development; there is also the need in mammals to ensure that there is also milk available for the new born and all the hormones mentioned have a part to play in mammary development and the ultimate secretion of this milk.

Nature is very prodigal in its provision to ensure the survival of a species. For instance, the heifer calf at birth has a potential of 75,000 eggs in her ovaries, but she will be very much above average if she produces as many as ten calves in her lifetime. The provision on the male side is incredibly generous, for instance one ejaculate from a healthy bull of 3–5 cc of semen will contain 3,000 to 5,000 million sperms, but only one of these can fertilise the egg that the cow produces at the end of her heat period.

ARTIFICIAL AIDS TO BREEDING

The most profound and far reaching advance that has been made in artificial breeding relates to insemination. With present knowledge and techniques of dilution, preservation, and impregnation, from 200 to 600 'doses' can be obtained from the one ejaculate of a healthy bull. A lot of semen from outstanding bulls, that potentially had an important part to play in herd improvement, was wasted until the development of the deep-freezing technique whereby 'straws' of semen, each one sufficient for one insemination, are preserved almost indefinitely, in liquid nitrogen at −196°C.

Artificial insemination has made its biggest impact in dairying, for apart from the fact that there has been a better sorting out of high-merit sires, there is the added advantage that milking cows are seen and handled at least twice a day and so it is relatively easy to detect the animals in season and thus take the necessary action. It is less easy with a suckler herd where animals may be seen only once a day. There are devices such as the use of a vasectomised bull fitted with a harness that carries a 'marker' pad but this does not remove the need for constant observation and the cutting out of cows from the herd as they come in season over a fairly protracted mating period.

Good progress, however, has been made in the control and synchronisation of oestrus by the use of synthetic substances with a similar chemical form to progesterone and the same function of suppressing heat. Most success has been obtained with pigs and sheep. In the latter species Robinson and his co-workers in Australia have developed a technique where polyurethane sponges, impregnated with progestagen (a synthetic progesterone), are placed in the vagina for 12–16 days. Two days after their removal the great majority of ewes will come in season together.

Unfortunately there has not been the same progress in the cattle field as with other species. Synchronisation is effective but usually conception rate to first synchronised heat is low. The most promising development in recent years is that involving the use of prostaglandins. These are biologically active unsaturated hydroxy-fatty acids. Injected non-surgically prostaglandins can induce very exact synchronisation within a few days with much better conception rates than those obtained by most other synchronising agents. At present the cost benefits of the prostaglandin treatment is uncertain, but technically the break-through is a very important one. Without effective synchronisation and a subsequent high conception rate, the effectiveness of AI in beef herds will never approach that of dairy herds.

INCREASE OF OVULATION RATE

It has been known for a long time that during early pregnancy in the mare there are large amounts of a hormone circulating in the blood which has a follicle stimulating function. It is known as PMS and it has been used experimentally by injection into the circulatory system to increase rate of ovulation in both sheep and cattle. In the early nineteen sixties there was sufficient confidence among those who were concerned with the development of this work in Britain to mount a field trial with dairy cows but it was not a success. Of the 416 cows that were treated, only 191 actually calved to the service following treatment. From these there were 35 sets of twins, 8 sets of triplets and one set of quintuplets. Ten per cent of the twin calves and 54 per cent of the triplets were born dead, and by two weeks these levels of mortality had increased to 16 per cent for twins and 71 per cent for triplets.

Altogether it was not a very happy experience either for the cows or for the farmers involved, and the Milk Marketing Board was perhaps unwise in being associated with this trial with the existing state of knowledge of the technique. The simple fact then, and now, is that there is insufficient knowledge about the optimal rate and timing of treatments. In any case, much as any dairy farmer would like to increase his sales of rearing calves, he can be 'penny wise and pound foolish' if he loses milk from his herd because of complications following multiple calvings, particularly if a high proportion of the calves are bad rearing prospects.

One can only conclude from this trial, and from subsequent field work in America, that though the technique has promise there is much to be done before induced super-ovulation becomes a practical proposition. Because of the difficulty of adjusting dosage to ensure twinning, rather than triplets or quads with their reduced chances of survival, interest is moving to the recovery of fertilised ova from treated cows and implanting them in host cows which, it is hoped, will produce two good calves. This still belongs to 'a brave new world' and so far as the practical farmer is concerned the best he can do at the present time is to work with nature and make the most of his cows by more conventional management practices.

IMPORTANCE OF PLANE OF NUTRITION

Of the management factors, one of the most important is the plane of nutrition of the cattle at point of service. If cows are losing weight up to service, conception rates will be impaired but with a rising plane there are good prospects that reasonable results will be obtained, always providing that bulls are fertile. Dr J. O. L. King, working in New Zealand with dairy cows, has published the figures given in Table 6.2 relating to conception rates and weight changes in the month prior to and including service.

These results are confirmed by experiences and experiments in

TABLE 6.2. EFFECT OF CHANGES IN BODY WEIGHT ON FERTILITY IN DAIRY COWS*

No. of cows	Gain in 4 weeks including service (kg)	Conception rate per cent
13	0·0	76·0
51	0·4 to 12·3	70·6
30	12·4 to 25·0	76·7
10	25·1 to 37·7	100·0
42	−0·5 to −12·3	28·6
35	−12·4 to −25·5	8·6
9	−25·6 to 37·7	0·0

* Quoted from Preston and Willis, *Intensive Beef Production*, Pergamon Press, 1970.

many parts of the world. Cows that calve late in the rainy season in the tropics seldom conceive in that year because the quality of grazing at the normal mating time is such as to prevent ovulation and the occurrence of heat and they become biennial breeders. In our own farming, cows that are subjected to multiple suckling are often very difficult to get in calf, presumably because of the effect of this on body condition.

It is inevitable that cows, and especially those that calve in high condition, will lose weight in the first few weeks after calving, but from a conception viewpoint it is important that weight losses are reversed from about the ninth week after calving. With calving just in advance of pasture growth, the average suckler cow should be in a positive nutritional balance by the time she is due for service but in an autumn-calving herd the position can be very different especially with the short duration of heat that normally occurs in the winter months. The farmer who practises autumn calving cannot afford to let his cows go to skin and bone before Christmas if he wishes to retain his intended pattern of calving.

CONTROL OF DISEASE

Condition of cows at various stages of the year is of such critical importance in relation to their performance that both the Milk Marketing Board and the MLC have developed systems where the animals can be given scores of body condition which are then related to what is considered to be optimal, taking into account economic considerations as well as actual physical performances. Details of condition scoring will be found in Chapter 12 where it is related to the management of suckler herds.

Cattle are subject to a wide range of disease conditions and physiological disorders which upset the normal processes of reproduction and though it is beyond the scope of this book to deal with veterinary aspects, emphasis must be placed on the need to maintain a healthy herd. A farmer's policy must be much more than one of calling in his veterinarian only when there is either a herd or an individual cow problem. There must be a management plan that will minimise trouble, and this is where the competent veterinarian can make his most productive contribution. The more livestock farmers use the veterinarian as a safety officer rather than as a fireman, the better it will be for farm profits.

Fortunately, after many years of effort by farmers and veterinarians alike with many disappointments on the way Great Britain has now what is hopefully regarded as a brucellosis-free national herd. There are still occasional outbreaks of brucellosis but fortunately these seem to be less frequent than they were a few years ago when some farmers had to endure repeated breakdowns in their efforts to create clean herds. The loss is not only that of premature calves but also the difficulty of getting affected cows in calf once more. Reproductive diseases other than brucellosis include trichomoniasis and vibriosis which are carried by the male. If a herd suffers from either of these problems it is preferable, if this is possible, to use artificial insemination in order to prevent cross infections. If a farmer is forced by circumstances to borrow a bull that has been used in another herd he is well advised to establish that it is clean in respect of venereal diseases for otherwise there is a danger of introducing something more than the bull into his herd.

There are other conditions affecting female reproductive performance, for instance retained cleansing which usually leads to the condition known as 'whites'. Then there is a variety of other infections of the reproductive tract such as vaginitis and metritis which are amenable to veterinary treatment.

Mineral deficiencies or imbalance can also weaken reproductive performance. Unfortunately there is still a lot of conjecture and very little precise knowledge about the role of minerals in this connection. It is known that deficiencies of phosphorus and excesses of calcium are associated with poor conception rates, but no one is yet in a position to state the optimal balance with any great certainty. Often a veterinarian will advise a farmer to offer a phosphorus-rich supplement where there is a breeding problem and cows are on a calcium-rich diet, but if this advice fails it does not constitute grounds for sacking one's vet, for he is doing his best with all the handicaps that come from limited knowledge of a very complex problem.

One last point requires emphasis though it is not a disease problem, namely calf losses at parturition. Most of foetal growth occurs in the last third of pregnancy, and gross overfeeding during this stage could lead to high calf weights and possible dystocia trouble. At the other end of the scale, underfeeding could impair the cow's body reserves and lead to reduced milk flow in early lactation. There is thus a need to preserve a happy medium in the nutritional treatment of cows as they approach calving.

Having said this, it is well established that different breeds of bull and cow have differing dystocia risks. Birth weight, sex of calf, gestation length and parity of the dam are all important factors. Surveys carried out in Britain clearly show greater dystocia risks from the larger sire breeds as can be seen in Table 6.3.

A similar situation is to be found in beef herds even though cows are generally on a much lower plane of nutrition prior to calving than they are in dairy herds. The MLC has been monitoring nine breeds in relation to calving problems in recorded suckler herds and the results of these surveys covering nine bull breeds are summarised in Table 6.4.

Again we see the greater risk of difficult calvings and higher neo-natal mortalities with the larger breeds as compared with the Aberdeen Angus, Hereford and Sussex. Nevertheless the higher

TABLE 6.3. DYSTOCIA PROBLEMS IN BRITISH HERDS FROM VARIOUS CROSSING SIRES (ALL CALVINGS FROM FRIESIAN DAMS EXCLUDING HEIFERS)*

Breed of sire	No. of bulls	No. of calvings	% serious dystocia	% calves dead
Charolais	29	4744	3·6	4·3
Chianina	10	1467	3·1	5·2
Hereford	62	10845	1·0	3·0
Limousin	6	845	2·4	3·2
Simmental (German)	13	3187	3·3	4·0
Simmental (Swiss)	10	2138	3·7	4·8

* Adapted from Wilson A et al 1976 Anim. Prod. 22:27–34.

TABLE 6.4. SUCKLER COW PRODUCTIVITY IN RELATION TO BREED OF SIRE

Sire breed	Assisted calvings %	Calf mortality %	Calving interval days	Annual prod. per cow kg
Charolais	9·0	4·8	374	208
Simmental	8·9	4·2	374	203
South Devon	8·7	4·0	375	203
Devon	6·4	2·6	373	200
Limousin	7·4	3·8	375	199
Lincoln Red	6·7	2·0	373	198
Sussex	4·5	1·5	372	196
Hereford	4·0	1·6	372	189
Aberdeen Angus	2·4	1·3	370	179

weaning weights of calves by bulls of the larger breeds compensates for the higher mortalities so that the Charolais, followed by the Simmental and the South Devon, tops the list in annual production per cow. Differences in calving intervals are small and are mainly attributable to the slightly longer gestation periods of the large breeds as compared with the Aberdeen Angus and Hereford.

There are insufficient data available to establish definitely whether a high incidence of difficult calvings is associated with a reduced herd life for suckler cows. The expectation is that this will be one of the hidden costs of using high growth bulls. In one MLC survey covering 19,000 calvings, in some 800 herds the Simmental and the Charolais both recorded 1·2 per cent incidence of surgical calvings as compared with 0·1 per cent for the Aberdeen Angus and 0·3 for the Hereford. In selection programmes the large breeds must include an emphasis on ease of calving, which is regarded as so important by AI organisations that the offending bulls are slaughtered.

Chapter 7

GROWTH AND DEVELOPMENT

ANIMAL GROWTH is a complex process which involves not only an increase in size but also changes in form and function of the different parts of the body. Growth begins at conception and continues until the animal achieves the mature size characteristic of the breed and species. The actual process of growth involves both cell multiplication and cell enlargement, together with the inclusion into the body cells of material taken from the diet such as calcium salts into bone and lipids into adipose tissue. Body cells become organised into tissues and these in turn make up the component parts of the animal body. All growth processes are under cellular and endocrinological control, but the actual growth achieved by any animal is a result of the inherent genetic potential and the environment under which it is living. As growth begins at conception the environment can be held to begin with that of the mother's uterus. Since growth has the components of size increase and also changes of shape, it is convenient to consider it in these two sections.

LIVEWEIGHT INCREASES

Most animal species have a typical growth curve that is sigmoid or S-shaped, involving a relatively slow period in early life, a self-accelerating phase and finally a self-inhibiting phase when growth again slows down and finally ceases. Changes in weight once mature size has been reached are merely due to the addition of corporal fat and do not constitute true growth. In general terms the termination of growth occurs when production of somato-trophic hormone (STH) from the anterior pituitary gland is insufficient to stimulate further growth activity. The amount of STH increases throughout life but expressed as the ratio of STH to body size it declines. Once an animal is mature, there is only sufficient STH to replace worn and damaged tissues and as a result no further growth occurs. Genetic differences between fast and slow growing animals may reflect their differing ability to produce STH.

Growth may, of course, be slowed down or even cease long before the animal reaches mature size. In this instance the limiting factor may well not be STH concentration but may be shortage of nutrients, poor housing conditions and the like. Adverse environmental conditions are often of greatest importance in early life, whereas in the later stage an animal is most frequently limited by its own internal physiology or by dietary aspects. This is one reason why products such as antibiotics (which protect against the environment) are most successful in early life, whereas growth stimulants which alter internal physiology (e.g. hormones) are most effective in older animals.

The stage of an animal's growth can be expressed in chronological or physiological terms. Thus two steers of a year old are at the same chronological age, but if growth has been delayed in one by virtue of poor nutrition it will be at an earlier physiological age than its partner. Physiological age can thus be expressed as the stage of maturity achieved. For example, Charolais and Jerseys will approach mature size at quite different rates and reach the same physiological age (i.e. same per cent of mature weight) at quite different chronological ages.

Liveweight growth is usually expressed in units of increase per day of life but this is subdivisible into various components. Increases in weight can be associated with tissue deposition, i.e. increases in muscle and bone and fat, or with changes in size of organs and viscera. Thirdly, increases in weight can be associated with gut fill. In cattle fed on low energy bulk feeds, a higher proportion of growth will be reflected in gut fill than would be the case in cattle fed cereal-based diets. Thus at the same liveweight the former would have greater gut fill and hence lower killing-out percentages than the latter. Differentials in growth rate would thus be more marked when expressed in terms of carcass growth than in terms of liveweight growth.

PATTERN OF GROWTH

All parts of the body compete for nutrients circulating in the blood stream and the priority of different tissues is governed by their metabolic rates. It has been established that tissues compete in the following order: brain and central nervous system, bone, muscle and fat. In the pregnant cow the foetus will compete with the cow's own tissues and have greater priority in early life, but this will fall as the foetus develops. In times of nutritional stress the different tissues will cease to grow in reverse order to their priorities. Thus a steer given inadequate feeding will cease to deposit fat. If further feed restriction occurs, muscle will also cease to grow and fat deposits will be utilised to assist growth of brain and bone which will continue to grow, albeit at decreasing rates.

Most cattle will spend some time when their growth is not so high as their genetic potential allows. With single-suckled calves on good land or with cereal-fed steers the period of restriction is negligible. Many other systems do result in prolonged periods of restriction and in those countries with pronounced dry seasons such restrictions can be severe. Cattle, in common with all other species, can exhibit compensatory growth when, after a period of feed restriction, they are fed again at a more normal rate. This phenomenon is characterised by a faster than average growth during the period of re-feeding. Some of this greater growth is due to increased gut contents but a high proportion is due to extra tissue growth which is laid down at a faster rate than is the case with comparable animals of similar chronological age which have not undergone feed restriction.

It is known that during compensatory growth the animal will have a greater feed intake than normal and convert feed more efficiently in liveweight and carcass terms. If re-feeding is carried on long enough, the animal will compensate entirely and end with a carcass which is more or less identical in compositional terms with that of normally fed cattle. If the period of restriction is extremely severe, such that it leads to losses of lean tissue or if it is carried out at a very critical stage of the young animal's life, then some lasting effects may be apparent. However, under normal British conditions this is unlikely.

Potential for compensation does, however, occur in most UK systems and can be of value to feeders if they are aware that an animal has been restricted. Such an animal will, at the fattening (re-feeding) stage, convert feed better than non-restricted cattle and also grow faster. It is thus an economic proposition to the finisher. He must, of course, be certain that the animal is one that has been restricted and not simply one which is of poor genetic potential. The economics of producing restricted cattle is another matter entirely since prices of suckled calves or stores will be related to weight and condition and hence restricted animals will command lower returns even though their subsequent perform-ance may be superior.

In integrated operations where cattle are taken from birth to slaughter under one ownership the economics of compensatory growth are less obvious. It can be justified only if there is a lifetime net gain in terms of feed required per unit of gain (preferably carcass gain). Where cheap feeds are available some case can be made out for it. In certain regimes in the UK compensation is an integral part of the operation and it behoves farmer and fattener to understand it if they are to maximise potential.

CHANGES IN FORM AND FUNCTION

The increase in liveweight of an animal is associated with a marked change in shape and function of various organs. At birth the head and legs are relatively large, but with time the upper muscles of the leg develop and the body becomes proportionately larger. Sir John Hammond who, with his Cambridge group, was a pioneer in growth studies, postulated that these changes in body proportions came about because different parts grew at different rates. This was true though a secondary claim that the most valuable parts of the carcass increased proportionally with age and nutritional plane is now known to be inaccurate. Subsequent work, mainly by Dr R. M. Butterfield in Queensland, showed that abdominal muscles were in fact the last to develop, growing at the relative rate of 135 compared with 100 for all muscles. The expensive muscles on the distal part of the hind leg grew at a relative rate of 90 while those surrounding the spinal column grew at the same rate as total muscle. This work proved that increasing size led to an ever-decreasing proportion of first-quality joints though in absolute terms these were bigger as the animal grew. The fattener must thus seek to kill at the weight that will maximise first quality proportions without being excessively fat, and yet still be large enough to provide a sizeable carcass.

Another of Butterfield's findings—that the proportion of expensive muscles was much the same in quite a variety of different breeds—came as a surprise to the traditional beef man, who has always claimed that the classical beef breeds were superior in this trait. There are, however, breed differences in respect of proportions of first quality meat when this is expressed as a percentage of total meat. This is illustrated by the data given in Table 7.1 which

TABLE 7.1. CARCASS ATTRIBUTES OF CROSS-BRED SUCKLED STEERS ACCORDING TO BREED OF SIRE

Sire breed	Killing-out %	Saleable meat in carcass %	Saleable meat in high-priced cuts %	Conformation score*
Aberdeen Angus	51·8	72·2	44·0	10·0
Charolais	53·8	72·3	44·8	11·1
Devon	51·8	71·4	44·0	9·0
Hereford	51·7	71·6	44·2	9·3
Limousin	54·0	73·0	45·4	11·1
Lincoln Red	51·6	70·8	44·2	8·6
Simmental	52·6	71·8	44·8	10·4
South Devon	52·6	71·6	44·2	8·4
Sussex	52·4	72·1	44·0	9·7

* 1–15 point scale with excellence at top end

have been derived from the MLC beef breed evaluation trials with suckled calves, the cross-bred progeny of 9 sire breeds, slaughtered at the same degree of finish.

The results establish clearly that the reputation that Limousin-cross cattle have earned of being good butchers' animals is thoroughly justified. Not only does it have the highest killing-out percentage but it also leads in percentage saleable meat and percentage high-priced cuts, and it has the same conformation score as the Charolais which compares more than favourably with the progeny of British breeds for the listed criteria.

Conformation of the superior carcass attributes of both Charolais and Simmental over British breeds is provided by other MLC data from dairy-bred cattle reared for slaughter at about 16 months of age at a standard degree of finish. These are given in Table 7.2 which also includes figures on daily rate of gain and age for slaughter. These results do not include any for the Limousin but nevertheless, the general picture is that the Continental and South Devon crosses have a higher daily rate of gain than cross-breds by the other tested breeds and they are also late maturing. Charolais and Aberdeen Angus crosses occupy the two extremes with, respectively, slaughter ages of 523 and 438 days and carcass weights of 261 and 179 kg.

TABLE 7.2. PRODUCTION AND CARCASS DATA FOR DAIRY-BRED CATTLE ACCORDING TO BREED OF SIRE

Sire breed	Daily gain (kg)	Age at slaughter (days)	Killing-out %	Carcass weight (kg)	Saleable meat %	High-priced cuts %	Carcass score
Aberdeen Angus	0·84	438	48·5	179	71·4	44·5	7·7
Charolais	0·94	523	51·5	261	71·6	44·9	8·8
Devon	0·86	440	48·5	186	70·7	44·6	6·9
Friesian	0·85	500	49·6	216	70·2	44·5	6·4
Hereford	0·85	453	49·1	195	70·8	44·5	7·8
Lincoln	0·84	486	50·5	212	70·5	44·0	8·4
Simmental	0·93	506	50·4	244	71·4	45·1	8·9
South Devon	0·90	491	50·1	227	71·4	44·2	7·7
Sussex	0·86	406	49·7	201	71·9	44·2	7·7

NB: High-priced cuts are a percentage of total saleable meat.

Management of these cattle undoubtedly affected killing-out percentages which are low as compared with those given in Table 7.1, but again the Charolais compares more than favourably with the other breed crosses not only in this respect but also with the Simmental for conformation score and percentage of saleable meat in the carcass.

Unquestionably the Continental breeds are making important

contributions to beef production in Britain and results such as these give an unequivocal answer to arguments advanced 20 years ago by vested interests opposing the importation of Continental breeds, in particular the Charolais.

EARLY AND LATE MATURITY

Later maturity in a growth sense applies to those breeds or animals in which the sequence of tissue deposition is more extended than in early maturing breeds. Nutritive priorities are for bone, muscle and fat in that order, hence these tissues are laid down at different rates. In an early-maturing breed such as an Angus the curves will be telescoped together to a greater extent than in say a Friesian, so that at any given chronological age or at any given liveweight the former will have a greater proportion of fat than the latter provided environmental conditions are the same. In general, early-maturing breeds are less suited to cereal beef systems, while later-maturing breeds may be less suited to extensive fattening operations. This comes about because nutritional plane also acts upon the rate at which different tissues are laid down.

The plane of nutrition affects both the overall rate of liveweight increase and the relative development of fat on the one hand and fat-free tissue on the other. Comparison of growth changes on a chronological age basis might confuse these two processes unless made on a fat-free basis. The classic Cambridge experiments on growth on different planes at different stages suffered from the final comparison of animals at the same age but different weights, or at similar weights but of dissimilar fat contents. More recent interpretations of these experiments have shown that plane has little effect on lean tissue production (relative to fat-free body) or upon bone content, but does lead to marked changes in fatness, this trait increasing as energy level of the diet increases.

In high energy diets (cereals), rates of change in carcass composition are accelerated such that an intrinsically early-maturing breed might become too fat at too early an age and have to be slaughtered at too light a weight. In contrast on a low plane diet a late-maturing breed might take too long to deposit the required amount of fat cover. This effect of nutrition on both the rate of body development and the differential effect of fat relative to fat-free tissues (muscle and bone) is a vital consideration in determining the kind of stock to use on any particular system.

SEX INFLUENCES

Sex influences will be of importance not only in respect of birth weights and subsequent growth, but also in regard to shape through secondary sex influences. It has, however, been shown that even such widely different animals as heifers, old cows and steers will have similar percentage of muscle when carcasses are examined at the same fat level.

The maternal environment will play its part and large cows can generally produce heavier birth weights than smaller ones, though nutrition in the final stages of pregnancy is probably most important in respect of birth weight and calving troubles. The lighter calves at birth are frequently born at an earlier physiological stage than heavier mates and may be more susceptible to environmental stress as a result. In a sense the heavier birth weight of the male is partly a result of a slightly longer gestation period and not merely its sex. Rapid growth after birth will not only affect compositional change but will also influence reproductive development. In general, sexual maturity is accelerated with higher nutritional planes and this, in turn, permits earlier breeding, although the efficacy of so doing must remain a question of the individual breeder's own economic situation. In breeding stock it is established that early restrictions may result in stunting of a temporary or permanent nature and will lower resistance to stress. In contrast, overfeeding is unlikely to lengthen the life of a breeding cow and could have the reverse effect.

Chapter 8

BASIC NUTRITIONAL REQUIREMENTS

LATER CHAPTERS deal with the practical aspects of various beef production systems in which the nutritional aspects vary considerably. Certain requirements are common to all cattle whatever their function; they need energy, protein, minerals and vitamins. The quantities required will depend upon function, i.e. whether the animal is growing, reproducing, milking or merely maintaining weight, and upon the kind of diet being used. In dealing with efficiency there are two aspects to consider—biological and economic. In the final analysis we are concerned with the economic efficiency, but it is essential to understand biological efficiency and it is with this latter aspect that we are concerned here. Biological efficiency can be defined as the maximisation of animal output in terms of meat, milk or calves from specific nutritional inputs.

ENERGY

Of the various nutrients energy is the most important; it constitutes the major item and hence the main cost of any diet. The ruminant is characterised by the amount of energy which it can absorb and metabolise in the form of volatile fatty acids (VFA). It is known that various mixtures of VFAs are used with different efficiencies, according to their relative proportions and to the function for which they are being used. Poorest biological efficiencies result from diets used for fattening which contain high proportions of acetic acid. In contrast, high efficiency is possible in maintenance diets producing low proportions of acetic acid. In general, one needs diets which yield low acetic levels and which are made up of soluble carbohydrates, so that these can pass through the rumen stage still fairly unfermented.

Various techniques for describing energy have existed varying from the now out-dated starch equivalents, through total digestible nutrients (TDN) to the new American system of Net Energy (NE). The most effective of these is probably the metabolisable energy (ME) system, which is the energy value of a feed after discounting losses in excreta, urine, as methane and as heat. The efficiency with which this ME is used for production (growth, lactation, etc.) will depend upon various factors, which include not only the actual production process but also the method of feeding, the chemical and physical nature of the feed and the age and body composition of the animal itself.

Diets which are in some way deficient in essential nutrients will be digested less efficiently than those which are properly balanced. Similarly, losses through faeces increase with the level of feeding above maintenance and inversely with the feed digestibility.

In fattening animals the efficiency of utilisation of ME increases in direct linear fashion with the ME of that feed. In other words, as the diet becomes less fibrous and more energetic the efficiency of its use for fattening rises. Thus at an ME of 2·0 Mcal/kg of dry matter biological efficiency will be about 40 per cent, but at 3·5 Mcal this will have risen to some 65 per cent. In the maintenance case, such as a non-pregnant reproductive animal, the range in efficiency from 2·0 to 3·5 Mcal would be from 65 per cent to some 78 per cent. In the lactating animal, efficiency also begins at a relatively high level and rises to a peak at about 2·8 Mcal whereafter it declines, not so much because overall efficiency is worse but because the animal starts to use feed for weight increases and hence milk producing efficiency is poorer.

In summary, one can conclude that biological efficiency for fattening gets better as the energy level of the diet increases and the fibre level declines. Biological efficiency for reproductive cows improves only fractionally with energy level, hence there is little economic point in increasing their feeding levels beyond a minimum.

There is some evidence that diets which lead to high propionic acid fermentations result in lower nitrogen retention than diets high in acetic acid production. Hence high concentrate diets lead to fatter carcasses than do high roughage diets, but despite this no clear evidence exists to show any relationship between VFA ratios and animal performance.

Of major importance is feed intake, since in crude terms the more an animal eats the more it will produce. Dry matter intake as an absolute value increases with animal size, but expressed as a percentage of body weight this falls gradually as the animal grows. On high energy diets the drop will be from about 3·0 per cent in a weaned calf of some 100 kg to 1·8 per cent in a 400 kg steer. When the diet contains some 10–20 per cent roughage, then dry matter intake will be some 0·2 percentage points higher. Similarly in cattle exhibiting compensatory growth, feed intake will be relatively high in relation to normal cattle (see Chapter 7).

It is generally accepted that increasing energy value of a feed leads to increasing energy intake, which will be maximised at about 70 per cent digestibility. Beyond this point dry-matter intake will decline and energy intake remain fairly constant. In high roughage diets, voluntary intake is largely controlled by the speed at which food passes through the digestive tract and by such factors as stage of fatness or pregnancy. At higher dietary energy levels the intake is basically determined by chemostatic or thermostatic factors. Thus a cow nursing two calves will eat more than one with a single, and a dairy cow will eat more in early lactation than late.

It is generally established that biological efficiency improves as concentrates replace roughages. Up to some 80–5 per cent concentrates there is a steady improvement in daily gain and a marked increase in feed efficiency, coupled with higher killing-out percentages due to less gut fill. If all-concentrate cereal diets are considered biological efficiency is maximised, but when some roughage is added (up to 10–20 per cent) there may be an increase in feed intake and some extra liveweight gain. Above that level both gain and intake decline as does efficiency.

The type of roughage used, whether it is straw or hay, seems to be of little importance in high-energy diets but the use of forms of synthetic roughage which at one time had some advocates is not recommended. The availability of roughage with high-energy diets, particularly where maize is used, gives protection against bloat and liver abscesses.

USE OF CEREALS

Almost all cereals can and have been used as cattle feeds though it does seem that wheat is less suitable at high levels than most others. Barley is the most widely used and has the advantage of containing husks that help to reduce the chances of digestive upsets. It is best fed rolled at about 16 per cent moisture or as a pellet when dry and unground. Grinding is undesirable as it destroys the husk structure and this is particularly disadvantageous when pelleted. Maize, on the other hand, has little fibre unless fed with the cob and can lead to more bloat problems. Sorghum is less satisfactory than barley though it may carry fewer bloat risks than maize. All three cereals have ME values of some 3·0 Mcal/kg with slightly higher values for maize with protein levels from 10 to 13 per cent. Other energy sources are molasses, though not used as such in Europe, beet pulp and whole beet. In the case of the latter two their value as replacement of grains is limited to an economic one since both will reduce efficiency and gain.

Various techniques exist for improving the efficiency of cereals

4. Charolais cross calves on rough grazing. The females in the picture are Hereford × Friesians.

5. Charolais-sired weaner calves out of Irish dams on a farm in Alnwick, Northumberland. This illustrates the need for a good dam to get the best out of Charolais.

6. Rearing beef calves on pasture. In this unit 17–18 nurse cows are used to rear groups of 50 sucklers which are brought in for finishing on concentrates.

of which the two most effective are heat processing and high moisture ensiling. Both can lead to improvements of 10 per cent or upwards in efficiency of feed utilisation, due to improved digestibility and to more rapid passage of digesta with resultant breakdown of starch post-rumen.

With grass diets the same situation does not hold as with cereal diets, since one is now dealing with a feed that has around 2·2 Mcal/kg and only about 24 per cent dry matter. Intake is more a question of gut fill than of VFA ratios, and energy intake may well be the factor limiting growth, since a 400 kg steer gaining 1 kg a day would need to eat 23 Mcals of ME daily, or about 40 kg of grass.

Clearly various situations exist regarding gains per day and energy contents of the diet and the reader can check these in specialised books on the subject.[1, 2]

Energy deficiency is one of the commonest deficiencies noted in cattle and will be demonstrated in reduction or cessation of growth, failure to conceive and lack of resistance to parasitic infections.

PROTEIN

After energy the major item to be considered is protein. Unlike the non-ruminant, cattle do not, in general terms, require proteins to be of specific kinds in terms of amino acid content. Moreover they are able to synthesise some of their protein needs through rumen micro-organisms, and thus have the capacity to utilise non-protein nitrogenous sources such as urea, which are of no value to species like pigs and poultry.

Protein tends to act upon growth directly rather than feed conversion efficiency and if given in excess will not have toxic effects but will simply be excreted and wasted. The feeder needs to ascertain the minimum level of protein necessary for health and the amount needed to maximise gain. He must then arrive at a protein level between these two limits which will maximise economic efficiency.

Protein deficiency will first be noted in depression of appetite and this will have an effect upon energy intake. As a result protein and energy deficiencies will frequently occur together. In breeding females deficiency may result in irregular or delayed oestrus, while in lactating animals milk yield may be affected adversely.

Cattle grazing good quality ryegrass will have no limitation in protein since energy intake would be the principal factor limiting growth. On high energy diets, however, there is no consistent

[1] The Nutrient Requirements of Farm Animals (No. 2 Ruminants), ARC, London, 1965.
[2] The Nutrient Requirements of Beef Cattle, NAS, Washington, 1970.

agreement between US and British standards as regards protein needs. It is usual to describe protein in terms of per cent of the dietary dry matter. Unfortunately, increasing the energy of a cereal diet will lead to corresponding reduction in dry-matter intake, and hence protein intake in absolute terms will tend to decline. In contrast, on high roughage diets increasing slightly the ME concentration will lead to increased feed intake and, at the same protein:energy ratio, to an increase in protein consumption. On cereal diets, therefore, the need is to think in terms of the amount of protein which will be consumed rather than the per cent of protein in the diet. Having said this, it can still be concluded that on cereal beef systems about 14 to 15 per cent protein is needed in dry matter for maximum growth and that this can be lowered to about 12–13 per cent when 10–20 per cent roughage is fed.

Urea can be used to substitute some true protein though it has only limited value in grass diets unless fed with cereal or other carbohydrate supplement. In cereal diets it can be used up to about 33 per cent of the protein requirements. This is not to imply that biological efficiency will benefit from such substitution, indeed a survey of the evidence leaves no doubt that urea substitution of protein will lead to reduced gain (about 6 per cent) and reduced feed efficiency (about 5 per cent). In economic terms, however, urea substitution of protein can be profitable since it can be that savings in feed costs more than offset reductions in gain and efficiency. Most other forms of NPN have similar results to urea and no evidence exists to show that less easily hydrolised forms of urea are more effective in stimulating growth. Some evidence exists that urea and animal fat may be somewhat incompatible in cereal diets and this needs to be borne in mind in feed preparation.

Since protein quality is of limited importance, protein sources can be used in relation to their known protein content and relative price. Some new evidence is becoming increasingly available to suggest that less soluble proteins, such as fish meal, which tend to pass through the rumen largely untouched, may be of better value in biological terms and this is particularly true in those diets using urea alongside true proteins.

This leads to a brief consideration of relative degradability of proteins from different sources due to the action of rumen microflora.* A protein which is soluble is also degradable, that is to say

---

* For further details consult the *Nutritional Requirements of Ruminant Livestock*, Agricultural Research Council (1980), Commonwealth Agricultural Bureau.

it is broken down by the microflora and then reconstituted as microbial protein which is then absorbed in the normal process of digestion. On the other hand, less soluble protein escapes from the rumen to be digested in the lower gut along with microbial protein. Where cattle are on little more than a maintenance regime and growing very slowly, degradable protein will suffice for their needs but at higher levels of production, for example, cows in full lactation or young fast growing cattle, then it is important to have a high proportion of undegradable protein in the diet. There is no problem on this account when cattle are on good pasture but performance on high-cereal diets and sometimes on silage diets can be improved by inclusion of fish meal or soya meal which both have a high content of undegradable protein.

MINERALS

Minerals constitute a relatively cheap part of the diet and as a result little conclusive evidence as to requirements exists. The ARC and NRC texts provide generally sound lines as to needs, as do most textbooks on nutrition. Over the normal fattening range it can be generally stated that some 0·20–0·28 per cent of dry matter are needed of calcium, phosphorus, sodium and chlorine, with about half of this of magnesium, and some 0·70 per cent of dry matter of potassium. The remaining minerals—iron, copper, zinc, magnesium, cobalt and iodine—are minor elements with requirements based in mg/kg of dry matter fed.

For general lines it can be safely assumed that grazing cattle will eat more salt than those fed dry feeds, and that salt intake will be higher as the forage becomes more succulent. Steers on hay will eat less than on silage and on high-cereal diets less than on high-forage feeds.

Calcium and phosphorus are closely related, though deficiency of the latter is more common. Since phosphorus is involved in energy metabolism, the shortage of this mineral will result in reduced gain following reduced feed intake. In the long term, bone development will be adversely affected, as can be reproductive processes. Most cereals do not require phosphorus supplementation but will need calcium to maintain satisfactory calcium: phosphorus ratios. Where heavy parasitic infection occurs, more phosphorus may be needed.

VITAMINS

Cattle have no need of supplementation of vitamins C and of the B complex since they can be synthesised directly. We are thus concerned only with A, D, E and K.

Under pasture conditions carotene will be supplied in the grass, and supplementation of vitamin A would not normally be needed in British conditions. With cereal diets, however, the situation is more complex and deficiency can lead to skin problems or blindness. It seems that at about 30,000 to 35,000 IU* per head daily the vitamin A requirements will be met, and further supplementation will not be needed. However, it does seem that on cereal diets some degradation of this vitamin occurs in the digestive tract, and it is normal practice to supplement at the rate of 6,000 IU/kg dietary dry matter. The feeder should, however, consider the previous history of the animal. Cattle can store vitamin A and carotene in the liver and in body fat during times of good pasture, and these supplies may be sufficient to maintain the animal for up to 6 months.

Vitamin D is generally believed to be unnecessary in cattle exposed to sunlight or eating sun-cured hay, but in cereal beef or artificially-reared calves some vitamin D may well be needed. This is usually given at the rate of about 25 per cent of the vitamin A requirements, i.e. 1,500 IU/kg feed dry matter.

Some American workers have established that vitamin E is now a necessity and the general requirement is at 0·6 IU/kg feed dry matter. A symptom of deficiency is muscular dystrophy and this can sometimes occur if there is heavy intake of nitrates.

There is no evidence that vitamin K is needed.

*IU = International Units

Chapter 9

REARING DAIRY-BRED CALVES

WITH DAIRY-BRED calves constituting the raw material for over two-thirds of the clean beef produced in Britain, rearing of these calves over the first 3 to 4 months of age is of very great importance, not only in terms of performance but also of cost. In this last connection the development in the mid-fifties of the early-weaning method of rearing and the substitution of whole milk by milk replacers has been of enormous importance. In addition to lower feeding costs, rearing of calves intended for beef can be undertaken in specialised units away from the farms of their birth and without any direct reliance on milking cows. This is to the liking of most dairy farmers who have enough pressure on their limited labour resources, attending to the needs of their milking and replacement stock, without having the additional burden of rearing calves surplus to herd maintenance needs.

Specialised calf rearing units are of two kinds. Some specialise in rearing from about a week old to the 12–14 week stage when they have reached a liveweight of about 100 kg, varying according to breed and sex, when they are disposed of to other farmers with the necessary resources for finishing them. Other fatteners prefer rearing their intake from the week-old stage themselves and this saves them the early rearer's margin of profit, but this necessitates an increased capital investment.

With both types of rearer there is usually a dependence on a linkman in the shape of a dealer who specialises in the purchase of calves at auctions on behalf of client rearers. Though the introduction of a middleman might be considered an additional expense it has, in practice, a lot to recommend it for many of the rearing units are some distance from the main dairying areas. For instance a Northumbrian farmer could easily spend an unwarranted amount of time at auctions in areas like Cheshire seeking his calf requirements and he could also be at the mercy of dealers who show their resentment of intruders by some judicious running up of calf prices. Buyers can be much better served by relying on competent dealers who understand the needs of their clientele and do their best to give satisfaction.

The movement of calves from breeding to feeding districts is now very well organised and is mainly done by insulated trucks. There are also safeguards to ensure that calves are in reasonable comfort and are given food in the course of long distance transport. The somewhat horrendous situation where in the middle of winter week-old calves born in west Wales travelled, by courtesy of British Railways to Aberdeen, a journey that would have stretched human endurance, even in a first class carriage, is now, very fortunately, a thing of the past. It has to be, for apart from animal welfare aspects, buyers cannot risk any hazard that endangers young calves that can cost well over a hundred pounds apiece landed on farms.

CALF LOSSES

Fortunately improvements in housing, hygiene and general management, including feeding and prophylactic measures, have greatly reduced calf losses which were substantial when week-old calves were cheap. Their enhanced value has undoubtedly concentrated rearers' minds on the aspects that really matter in maintaining thrift and minimising mortalities. At the breeders' end there are inevitable losses, either still-born calves or weakly calves that succumb soon after birth. These losses, which can be as high as 5–6 per cent, are undoubtedly aggravated by calving cows in high condition, the so-called steaming up to enhance milk yields. On top of these losses there are the subsequent rearing losses, which for the sake of profit in an operation where margins are invariably low, must obviously be kept to a minimum. If these exceed 5 per cent there is cause for real concern and in fact 3 per cent is now about the acceptable limit to mortalities in a good commercial unit. Losses tend to vary with season, the highest incidence being in the late winter–early spring period and the lowest in the late summer–autumn period. This seems to be related to the gradual build up of diseases in calf houses following prolonged usage and a reduced level of infection following resting in the summer when few calves are born. There is a moral to this observation namely the importance of giving calf rearing houses a thorough cleaning and disinfection and a period of rest between intakes of calves.

A key factor in calf thrift, which is a responsibility of the dairy farmer not the subsequent rearer, is making certain that they receive colostrum not later than 36 hours after birth because the intestine gradually becomes impermeable to the transfer of the antibodies. The dam's colostrum is a rich source of fat soluble vitamins, which are deficient in the new-born calf as well as the antibodies that give protection, particularly against *Escherichia coli* infections which constitute one of the major problems in calf rearing. Work at the National Institute for Research in Dairying (NIRD) has shown that *E. coli* infections arise not only through a lack of colostrum but also through localised infections in the intestine. It is a highly ubiquitous organism and it can multiply very quickly to cause such severe scouring that deaths occur as a result of dehydration.

Feed can be implicated in *E. coli* problems for it seems there is a greater build up of the organism with milk substitutes than with whole milk and this is believed to be related to milk processing methods. Most substitutes are based on dried skimmed milk which can be produced in different ways. Milk that is subjected to prolonged high temperature is associated with poor calf performance. The best kinds of dried milk are those where there is minimal denaturation of the proteins so that there is unimpaired digestibility. For instance milk subjected to pasteurisation for short periods (e.g. 77°C for 15 seconds) followed by spray drying is very satisfactory because the clotting function of rennet is not impaired as it is in high temperature dried milk and this reduces the risk of *E. coli* infections.

METHODS OF MILK FEEDING

There are several styles of milk feeding, each of which has its practitioners. The traditional method is twice-a-day bucket feeding with the calves receiving a prescribed ration of warmed milk until about the 4–5 week stage when the calves are abruptly weaned from milk and have to rely entirely on drinking water and concentrates with some hay which has been available *ad lib* from the outset of rearing. There is now a school of thought that favours once-a-day milk feeding, again with restricted amounts and another that supports *ad lib* feeding of acidified cold milk from milk dispensers for the first 3 weeks followed by gradual weaning over the next fortnight when milk feeding ends completely.

The *ad lib* system was initially developed mainly for veal production and was based on dispensing warm milk via teats which makes it necessary to train calves in their use. This normally takes about a couple of days. There has been a move away from warm to cold milk which, through acidification, is kept fresh for 2 to 3 days and here with dispenser feeding the calves are usually kept in groups with no more than 6 calves to each teat. The MLC has collected farm data enabling a comparison of *ad lib* dispenser feeding for 3 weeks followed by a gradual weaning over a fortnight with controlled bucket feeding of milk. There was virtually no difference in mortality rates between the 2 systems but the *ad lib* calves at 14 weeks, after approximately 12 weeks on their feeding regime, averaged 5 kg more than the bucket-fed calves but this had a high cost in terms of milk replacer, 26 kg as opposed to 12 kg per calf. The bucket-fed calves ate 170 kg of concentrates as opposed to 164 kg for the dispenser-fed calves but the overall food costs for the latter group were 30 per cent higher than those for the bucket-fed calves. Since one would expect a difference as small as 5 kg in liveweight at 14 weeks to mean little after a few months on the same diet there does not seem to be any advantage in *ad lib* dispenser feeding unless it is in respect of labour economies.

The bucket-fed calves in the above study combined both once-a-day and twice-a-day feeding. It has been shown there is little difference between the performances of calves on the 2 systems but with once-a-day there is a useful saving of labour and this is the method that the MLC recommends. Certainly with once-a-day milk feeding there is a greater likelihood that buckets and other utensils will be kept in a cleaner condition than they would be with twice-daily usage of utensils.

MANAGEMENT TO 14 WEEKS

The following regime is suggested for rearing Friesian and Friesian-cross calves up to 12–14 weeks.

(a) All calves should be inspected on arrival and any showing ill health should either be rejected or isolated from the main body. It is not now possible to use antibiotics as a protection against disease but it is advantageous to give them a multi-vitamin injection immediately after their arrival. They should be bedded down on clean straw in accommodation that has been thoroughly cleaned and disinfected. As a further safeguard against the inadvertent introduction of disease it is advisable for the reception quarters to be separated from other calves in the unit, with calves not being moved to the main body in the unit until it is abundantly clear that they are free from disease.

(b) Some rearers prefer to give calves about 250 g of glucose in 2–3 litres of water in 2 feeds on the day of their arrival but more usually calves go immediately on to milk replacer which is reconstituted according to the manufacturer's directions. Commonly this is 150 g of replacer (17–20 per cent fat in dry matter) to a litre of warm water. Feeding of this mix will start at a litre per day but over the first week it is gradually increased to its maximum rate of 2½ litres daily given in either 1 or 2 feeds, according to choice.

(c) It is advisable to pen calves separately while they are on milk as this is a safeguard against ear and navel sucking, but the animal welfare code insists that penned calves should be able to see other calves.

(d) Early weaning concentrate, either a proprietary brand or a home mix, such as that given in Table 9.1, should be on unrestricted offer from the outset along with some good soft hay. Apparent consumption of the latter may appear to be small but nevertheless it is necessary to promote rumination.

(e) Calves may be weaned abruptly when they are eating a kg of concentrate daily. This is normally at 4–5 weeks of age. Once the calves have settled to a diet without milk they can be bunched in groups of 5–10 matching calves and changed from the relatively expensive early-weaning mixture to a cheaper concentrate such as that given in Table 9.1. Concentrate consumption increases rapidly with weaning, and especially after bunching because of the social nature of feeding. When it reaches 2·5 kg per day it can then be maintained at this limit with any residual appetite being satisfied by an additional intake of hay. Good quality silage can be substituted for hay at this point.

(f) Such operations as dehorning and castration are best undertaken at 3–4 weeks of age when the calves are penned separately, with the proviso that they should be in good health.

(g) Feeding regimes after 12 weeks will depend on the time of year and the purpose calves will fulfil. If they are intended for cereal beef, and in this case they should be entires rather than castrates, they will go straight on to an appropriate cereal diet. Where there is a growing period in the interim before going out to pasture they will continue on hay or silage with some concentrate supplement. At this stage it is reasonable to expect a liveweight increase of 0·8 kg per day.

HOUSING AND DISEASE

Probably the biggest single cause of loss in rearing calves is enzootic pneumonia which is associated with a combination of inadequate ventilation, excessive condensation and wet bedding. Condensation is invariably a problem where there is bad ventilation but ventilation can also lead to draughty conditions unless precautions are taken. The risk of pneumonia is greatest in the still, humid weather characteristic of November and December which is the height of the calf rearing season. A building with a low ceiling is invariably much warmer than one with a high ceiling and in one respect this is an advantage but in another it can be detrimental because of an inadequate supply of fresh air. Under these conditions it is advisable to use fans to assist air movement. Movement of air at calf level can result in draughts and because of this, solid divisions between pens are recommended with the animal welfare proviso that individuals are not in visual isolation.

Calves can be reared very successfully in demountable pens with solid partitions erected in general purpose buildings. A cosy, draught-free lying area can be created with straw as part cover over the pens. After use the pens can be taken down for cleaning and disinfection before being stored for subsequent re-use, so that the building is left free for other uses.

Regardless of the type of accommodation it is important to have an isolation unit for sick animals because disease has an unfortunate tendency to multiply. At the first hint of a serious spread of disease it is important to get veterinary advice because it may be necessary to use antibiotics and these can only be used under veterinary supervision.

Possibly the most devastating of all the disease problems that can hit a calf rearing unit is a *Salmonella* infection, for here losses due to mortality are aggravated by a serious loss of thrift in survivors that require many months for recovery. Immediately there is any suspicion that scouring is not due to a dietary upset it is important to get veterinary advice for once *Salmonella* gets a hold in a unit it is hard to eliminate because the infection can be carried by rats.

CONCENTRATE FEEDING

A very high proportion of hand-reared calves in Britain are fed proprietary concentrates usually in the form of pellets with a high protein formulation for the first 5–6 weeks and a lower protein content for the later stages of rearing. Many farmers however, may prefer to use home-mixed concentrates and specimen formulations for this purpose are given in Table 9.1

The early-weaning mixture which has, in part, to take the place of cow's milk must have a high content of digestible protein. It is considered that 60–80 g of protein per Mcal of metabolisable energy is about optimal. The protein in white fish meal has a high biological value and for this reason it is an appropriate constituent of the diet before full ruminant functions are developed. Palatability is another important attribute and this is enhanced by the inclusion of flaked maize and molassine meal. The latter has another useful function, namely it reduces the dustiness of the ration. Rolling and bruising of cereal ingredients rather than grinding is stipulated for this reason while the coarser structure of the mix seems to be more acceptable than finely ground material.

When the change is made from one mixture to another one should avoid making a sudden break. If the 2 mixes are incor-

TABLE 9.1. CONCENTRATE MIXTURES IN CALF REARING (PER CENT)

Ingredient	Early weaning	Weaning 12 weeks	After 12 weeks
Rolled barley	—	75	80
Flaked maize	40	10	10
Bruised oats	35·3	—	—
Molassine meal	13	—	—
White fish meal	11	9	4
Soya bean meal	—	5	5
Salt	0·4	5	5
Limestone	0·3	—	—
Mineral/vitamin mix*	—	1	1
Vitamin A	6 million IU per ton		
Vitamin D	1·5 million IU per ton		

* Vit D3, Cu, P, NaCl, Fe, Mn

porated in the feeding bucket or trough for a few days there should be no serious drop in intake. The supplement fed after 12 weeks should contain about 13–14 per cent of crude protein unless really high quality silage is on offer. Under these circumstances the fish meal could be left out of the mix and the soya meal component increased to 9 per cent in order to cheapen the ration.

PRODUCTION TARGETS

MLC farm surveys, as well as experimental and demonstrational trials at the National Agricultural Centre at Stoneleigh, have made it possible to establish realistic production targets for both controlled feeding of milk replacements and *ad lib* feeding of acidified cold milk from dispensers. These are given in Table 9.2.

TABLE 9.2. REARING TARGETS FOR FRIESIAN AND HEREFORD × FRIESIAN CALVES

	Once daily (bucket)	Ad lib (acidified cold milk)
Daily gain (kg)		
up to 5 weeks	0·5	0·7
5–12 weeks	0·9	0·9
overall	0·7	0·8
Weight (kg)		
start	45	45
5 weeks	60	70
12 weeks	105	115
Feed (kg)		
milk replacer	13	25
Concentrates		
up to 5 weeks	30	10
5–12 weeks	130	140
Total concentrates	160	150

Obviously the calf mortality target is a clean sheet, but some deaths are inevitable and the average mortality for 25 calf rearing groups surveyed by the MLC in 1982 was 4 per cent. However there were no mortalities in the top-third of the sample judged on gross margin per head. This group also achieved production standards approximating to those given in Table 9.2.

ALTERNATIVE REARING METHODS

Because whole milk has long been an expensive commodity in Britain as well as other parts of the EEC, the emphasis in calf rearing over the past 30 years has been that described above, namely the use of milk replacers with concentrate supplements. This could change with imposed restrictions on milk output and there may be a return to a greater reliance on whole milk as there is in many countries abroad, where dairying is a less sophisticated operation than in Britain. For instance it is a common practice in some Central and South American countries to adopt a system of 'share' milking, where calves are able to suckle their mothers by day but are shut away from them overnight so that the cows can be milked the following morning to cater for human needs. It may seem rather a primitive form of animal production but neverthe-less it is very appropriate to tropical conditions where hand-rearing of calves is often a very precarious undertaking with high rates of mortality. Under this form of management the calves at least are assured of getting some milk, which is their natural food, without the hazard of the unhygienic conditions so often associated with bucket feeding in the tropics.

New Zealand research workers developed an interesting tech-nique for dairy farmers where heifer calves intended for herd replacement, up to 4 at a time, were allowed to suckle a nurse cow up to 6 weeks of age. They were then weaned and the cows were returned to the herd where their subsequent yields compared very favourably with the total yield of cows that had not been used for suckling. It was suggested that the suckling, combined with complete evacuation of the udder in early lactation, could have a stimulating effect on yield.

The use of nurse cows for multiple suckling of up to 4 calves at a time for a period of approximately 3 months after which they are placed by another smaller group for a similar period, had some popularity in Great Britain prior to the development of early weaning, which is a much more attractive system because it is more flexible and requires much less labour. Moreover, with up to 3 lots of calves being reared on each cow over the course of the year there is a recurrent procurement problem as well as calves of assorted ages and sizes. One of the great virtues of the early-weaning system is that procurement can be a short duration operation while calves in the course of rearing will usually be in comparatively uniform lots. Nevertheless with milk sales restricted by quotas some dairy farmers may decide to use 'problem' cows, eg. slow milkers and fractious animals, for multiple suckling, not on a successive batch system but with one lot of up to 4 calves over their entire lactation.

Finally a word about double suckling where a suckler cow with a good flow of milk, such as one would expect from a mature Hereford × Friesian, is required to rear a fostered calf as well as their own. On paper it seems an attractive proposition because the overheads of maintaining the cow are more widely spread but in practice there are snags. First, good calves are expensive and prices tend to be at a peak in the spring which is also the height of the suckler-calving system. There is a risk of buying infection along with the calf which can be transferred to the cow's own calf. A great deal of time and patience is required to ensure that the foster calf is accepted by the cow that naturally favours its own calf and this could be at the expense of other more important farm work. There is some advantage to be gained in mothering-up from linking 2 calves together with a short length of swivelled chain (40–45 cms) attached to leather collars when the calves, after a period of controlled suckling, are turned out with the cow. The chain is removed when the second calf is completely accepted. A cow suckled by 2 calves will produce more milk than it will with only one calf but the increase will not be sufficient to ensure that the cow's own calf will be weaned in as good a condition as it would have if it were the sole beneficiary of its mother's milk. Possibly double suckling has merit where a deep milking cow loses her own calf and the task of mothering-up will not be much greater with 2 purchasd calves than it is with one, but it is not recommended as a whole herd operation.

Chapter 10

BULL BEEF AND HORMONE ADMINISTRATION

APART FROM beef produced from cull cows and bulls and surplus heifers the beef industry in Britain uses the steer as its major production unit. In this respect the UK follows the same pattern as North America and most other English-speaking countries. In continental Europe insistence upon a castrated animal has never been as pronounced as in Britain and since the Second World War beef from entire males has been of increasing importance. The same is true of most developing areas of the world except where their beef is intended for British markets. Castration was the usual policy in New Zealand until the development of an export trade in beef to the United States. There the requirement is low-fat, boneless beef suitable for the manufacture of hamburgers and bulls are preferred to steers for this purpose.

BULLS VERSUS STEERS

The use of steers for meat production is of some antiquity and in its early days was probably based upon sound reasoning. In the past cattle were used largely for draught purposes and castration would undoubtedly render the animals more tractable. Added to this, the general nutritional plane of cattle was low thus leading to relatively lean carcasses whereas castration would tend to increase fat at a time when man's need was for considerable quantities of fat in his diet. Slaughter age of cattle was also much higher than it is in modern times; castration would have helped in the management of cattle kept on free range, as well as reducing the risks of indiscriminate mating which became important upon the setting up of clearly defined breeds of cattle and the early attempts to introduce selective breeding.

No one disputes the wisdom of man's decision to castrate those bulls not needed for breeding purposes at a time when farming was a much more primitive operation than it is today, but there is no justification for the blind acceptance of tradition for tradition's sake. This is of particular importance when it is considered that many of the early reasons for castration no longer hold water.

114

Cattle are not used as draught animals in developed countries and even among the developing nations their utility in this aspect is declining rapidly. Similarly, land enclosure and the increasing effectiveness of fencing has made management of non-castrates easier, while improved standards of animal nutrition and general management have reduced the life expectancy of the animal destined for beef such that it is now generally slaughtered at no more than 2 or 2½ years. Finally, the earlier requirement for fat in the human diet has been drastically reduced. Whether or not animal fat is linked to heart disease and other disorders the fact remains that the modern housewife, whatever her native tongue, is increasingly seeking meat that carries with it the minimum of fat. One is thus faced with the inevitable conclusion that castration is practised mainly because our grandfathers did so and there has been little incentive to question tradition.

BULL BEEF IN BRITAIN

There has been a history of official discouragement of bull beef in Britain which, in its entirety, represents a classic example of bureaucratic restraint on technical and economic progress. Only comparatively recently has it been possible for the producer to feed bulls rather than steers if that is his preferred course of action. Prior to entry into the EEC the Ministry of Agriculture, as agents for official policy, maintained a two-pronged system of discouragement. The first was a stipulation that all bulls must be licensed for breeding or castrated before 10 months of age which was part of an outmoded and quite unsound policy of livestock improvement that many believed was only maintained because of the vested interest of the people concerned in its administration. There was some relaxation because of growing pressure and a special licence to rear bulls for beef was available, but at a disproportionately high cost. The Ministry's citadel however, was not completely breached with this concession for at that time there was a rearing subsidy paid on potential beef calves at 9 months of age. In the case of bulls this subsidy was not paid until slaughter and then only if the carcass met with the approval of the certifying officer who, as often as not, was suspicious of, if not prejudiced against bull beef. The farmer who submitted bull beef for certification on a dead weight basis could find himself out of pocket on two counts; the loss of the rearing subsidy and the loss of the deficiency payment that was part of the guaranteed price.

With Britain's membership of the EEC, where bull beef is generally acceptable, there has been some relaxation of the old restraints. There is no really serious impediment with yarded bulls destined for early slaughter, eg. barley beef, but the use of bulls in the 18-month system is not so straightforward because the safety code for grazing bulls requires standards of boundary fencing that are expensive to the point of being a discouragement. On top of this there is the added complication of footpaths and rights of way which are so jealously guarded in Britain that any infringement of ramblers' rights will create problems. Even apart from any risk the farmer or his men run from handling older bulls, either in the fields or in the yards, which is by no means negligible with bulls of dairy breeding, it may sometimes be preferable to castrate males and use implants to simulate the leaner and more rapid growth pattern of the entire.

Castration of single-suckled males is general not only on grounds of tractability but also because of the risks of precocious matings, since it is not uncommon for heifers as young as 6 or 7 months of age to steal the bull and well-grown bull calves can become sexually mature at a similar age. The risks of unwanted pregnancies can be reduced by separating cows with bull calves from those with heifers but this is a further complication in management which does not appear to be justified until such time as there is preferential demand from finishers for entire weaners.

THE CASE FOR BULL BEEF

There is now a substantial volume of scientific literature as well as a great deal of farm experience in many parts of the world on the relative merits of castrate and entire males for beef producers. As might be expected, most of this is of European origin but there has been a growing increase in data from North America, Eire and the United Kingdom. This chapter is intended only as a review of the situation since extensive summaries already exist elsewhere.[*]

There seems to be general agreement among most scientific studies that daily gain is greater in bulls than steers when reared under the same conditions. The extent of bull superiority in this trait varies according to the specific circumstances but, in general, ranges from 5 to 20 per cent with about a third of cases falling in the 5–10 per cent range of superiority and over 40 per cent within the 10–20 per cent superiority range. Thus a fattener getting an average of 1 kg per day from steers could expect to push this up anywhere from 1·05 to 1·2 kg were he to change to bulls. These figures refer to animals fattened from a traditional weaning weight

---

[*] *Meat Production from Entire Male Animals* (D. N. Rhodes ed.), Churchill, London, 1969.

of about 200 kg. If one was to start with calves then superiority of bulls would be less apparent in the early stages of fattening since androgenic hormones do not become important until puberty is reached. Similarly, there is some suggestion that the plane of nutrition in energy terms is of importance.

Under tropical pasture conditions, such as exist in parts of South America and Africa, bulls have not always demonstrated any advantage in growth rate over steers and in some trials steers have out-gained their male counterparts. In such circumstances castration is probably to be preferred, particularly if the carcasses are destined for export and some degree of fat cover is needed to aid carcass preservation during transport. This is not, however, likely to be the case under most British conditions where both grass growth and quality are better than that of tropical areas.

Experiments undertaken in Eire using animals fattened on pasture have shown marked advantages in growth rate, up to 15 per cent for bulls and these values have been substantiated in Sweden. Despite this it does seem that the advantages of bulls, in growth rate terms, will tend to bear some relation to diet and that the more intensive the dietary system is in energetic terms, the greater will be the superiority of the bull.

Sex advantages are not, however, restricted simply to those of daily gain but are allied to benefits in improved feed conversion efficiency. Experimental work on this aspect is less well documented in view of the difficulties of measuring efficiency with pasture-fed cattle, but such data as exist are conclusively in favour of bulls. Superiority has depended upon the growth rate responses such that the greater the differences in daily gain, the greater the advantage of bulls in feed efficiency. In general, however, values have been in the range of 10 per cent higher efficiency for entire males.

A major advantage of the better feed utilisation of bulls is that they may be kept to larger weights than steers without adverse effect on feed costs. In general terms, bulls can be slaughtered at higher weights than steers for the same financial outlay. This will, in turn, have an effect upon carcass traits and it is here that the major advantages of bull beef are to be seen.

CARCASS QUALITY OF BULLS

There is unanimous agreement that any given weight bulls will have a higher percentage of edible meat and a lower percentage of fat than steers. Bone content will be fairly equal or be fractionally higher (1–2 units) in bulls. Since carcass fatness is related to killing-out percentage there is a tendency for steers to kill-out higher than bulls when slaughtered at similar weights. This advantage is offset by the greater lean content of bull carcasses, and, since bulls may be fattened to higher weights, the disadvantage in killing-out would be largely offset since this trait is also positively related to liveweight.

Not only do bulls tend to give higher values of edible meat but a higher proportion of this tends to be in the higher priced cuts. Despite the different shape of bulls and steers American and British workers have shown that the former yield more first quality meat and that this advantage is increased when trimmed carcass (ie. after removing excess fat) is considered. The leaner nature of bull carcasses is a reflection of their faster growth and greater feed efficiency since it is well established that fat is more expensive to produce than muscle. Some Nebraskan work has indicated that steers consumed almost twice as much energy (8·32 Mcal Digestible Energy vs 4·57) per kg of edible product. Allied to their greater leanness, bulls will yield greater rib-eye areas both in absolute terms and in terms of area per unit of carcass weight. Protein percentage of the meat will also be fractionally higher in bulls.

Although there is no evidence that castration was ever practised as a means of affecting meat quality, exponents of castration might point out improved meat quality in steers. In some aspects there is justification for this viewpoint although it has usually been exaggerated out of all context. Under pasture conditions it is possible that bulls will produce darker-cutting beef than steers, but there is no evidence to suggest that this will be so pronounced as to render the meat unacceptable. Under cereal beef situations, with bulls killed at 12–18 months, there is unlikely to be any recognisable effect on lean coloration. Clearly when comparisons are made between old breeding bulls the entire male is at a disadvantage in almost all aspects of meat quality but we are concerned here with the bull reared as a meat animal—not bull meat as the aftermath of a long breeding life.

COMPARING MEAT QUALITY

Aside from the tendency to darker coloration, no clearly defined situation exists in respect of most other traits. There is some evidence that bull meat may be more coarsely textured and also less tender. Tenderness is the trait most readily evaluated by the consumer, but not all studies have demonstrated effects of sex upon tenderness. Nevertheless, when such effects have been noticed they have always favoured the steer, whether undertaken with shear machines or by taste panel studies. Since tenderness is closely related to age, it is obvious that the age of the animals studied will be a vital consideration and it does seem that the adverse age effects upon tenderness are more pronounced in bulls than steers. The Meat Research Institute has claimed that bull meat is more resistant to compression and requires a greater effort to bite through it, but that this difference is most noticeable at higher cooking temperatures. The MRI suggested that when stewing bull meat it would require somewhat longer cooking than that from steers, but that with roasting differences would be minimised when meat was rare rather than well-done.

Juiciness is not an easy trait to evaluate, but American work indicates that sex differences are not particularly apparent in animals slaughtered under some 600 days of age. Above that age juiciness may be more adversely affected in bulls. A similar effect is noted with flavour and indicates that age is a major factor to consider in bull beef production since the closer animals approach two years of age, the greater will be the chances of deleterious traits being observed in entire males.

Differences in meat quality do not, of course, imply that bull meat is less acceptable to the general public. Consumer acceptance studies undertaken in Britain and North America, as well as in parts of Europe, have not shown any consistent prejudice against bull meat; it may be that some butcher antagonism is of a somewhat personal nature related to a desire to find some way of downgrading a carcass rather than bearing any relation to cut-out value. Several years ago when bull beef was under criticism from the trade, importations of bull beef from Yugoslavia met with a very firm demand, especially from firms that catered for such institutions as schools and hospitals where fat meat is not appreciated. It is a feature of trading interests that a commodity, provided it is morally respectable, becomes acceptable if it is possible to make a good profit handling it. Seldom does one now hear complaints about bull beef provided it comes from animals under 18 months of age and the resulting carcasses have been properly conditioned.

In conclusion one may assume that bull beef has much to recommend it from both efficiency of production and in terms of maximising lean meat production. The benefits of bull beef will be greatest in the more intensive systems although they will still be measurable in those less intensive systems with slaughter up to 18 months of age. Bull beef production would doubtless benefit those breeds such as the Angus which tend to deposit fat early and which, if left entire, could be fattened to higher weights without adverse effect on carcass traits or production costs. Those breeds with a tendency to low meat:bone ratios may be aided most by castration unless fattened in cereal beef operations.

PARTIAL CASTRATION AND FEMALES

Some years ago Russian scientists advocated a system of partial castration, while in recent years Australian researchers have turned their attention to the artificial cryptorchid. Both systems seem to have much in common and in general terms it can be concluded that they will produce growth rates, feed efficiencies and carcass traits that fall intermediate between steers and bulls. The future of such practices is limited, particularly where they call for greater skill than simple castration.

There are more than half a million beef-cross heifers slaughtered each year. In general females grow more slowly than steers and produce fatter carcasses. The differences are so pronounced as to make it necessary to slaughter females at much lower liveweights or to feed them on lower energy diets than steers if carcasses are not to become too fat.

GROWTH PROMOTERS

These are of two types, anti-microbial growth promoters which are administered as food additives and hormones which are usually administered as subcutaneous implants, or which can be put in feed. The anti-microbial agents are more important in pig production than they are in cattle production where in 1983 there were only 2 substances—Monensin sodium and Flavomycin—that were licensed for use. This position is likely to change because several other compounds were then under test for use with cattle.

The mode of action of these compounds has not been clearly established and there is little known about their possible effects on either carcass composition or muscular tissue. Monensin changes the microbial population of the rumen to give an increase in propionic acid production which in its turn results in improved feed conversion. The MLC estimate that about 50 per cent of the proprietary compound food for beef cattle and 70 per cent of concentrated protein supplements contain an anti-microbial growth promoter. There do not appear to be any problems of residues remaining in the carcass but always with substances affecting bacterial populations there is a risk of resistant strains developing and some of these could even constitute a danger to human health.

HORMONES IN BEEF PRODUCTION

Synthetic steroids were first used in beef production by workers at Purdue University in the United States in the early 1950s. Substantial improvements in growth rate and efficiency of food utilisation were secured and following publication of results their use became so widespread in the United States that it was estimated that by the 1970s three-quarters of the feeding cattle were being given diethyl-stilboestrol (DES) with a saving of £60 million annually in feed costs. Their use was also taken up in Britain but here the preferred hormone was hexoestrol which was considered to be safer because of its lower oestrogenic activity.

These hormones can be administered orally as food additives or as an implant at the base of the ear. The second method was the preferred option in Britain and the ear was chosen as the site of implantation because it lessened the risk of any residue in the dressed carcass. There have been a wide range of reported improvements in growth rates and food economy as a result of hormone administration. They appear to be greater in high cereal diets, such as that in barley beef production, than they are in diets with a high proportion of roughage or where cattle are grazing pasture. Gains in growth rate of the 14–16 per cent and 7–10 per cent in efficiency of food are fairly typical of reported results.

The mode of action of these hormones is not well understood. The improvement in food efficiency is probably attributable in part to an increase in lean meat production at the expense of fat deposition which has high energy demands. There are side effects from their administration which are more pronounced with DES than they are with hexoestrol. In the steer these consist of increased sexual activity, depression of the loin due to loosening of the muscles, a high tail end and some mammary development. However the side effects in heifers are much more pronounced, for instance vaginal prolapse and increased mammary developments to the point that these synthetic oestrogens have had no place in finishing heifers. A rather different heifer response is obtained with the administration of synthetic progesterones which are of the same nature as the oral birth control agents used by women. They eliminate oestrus during the period of their use and do not appear to have any harmful side effects. Gains of the order of 11 per cent in growth rate and 8 per cent in feed efficiency have been reported in trials comparing treated and untreated heifers on finishing diets.

MISGIVINGS ON THE USE OF SYNTHETIC HORMONES

There has been a lot of disquiet in many parts of the world concerning the administration of synthetic hormones to cattle intended for slaughter because of their alleged carcinogenic effects even though there is no evidence of residues in the meat of treated animals. Nevertheless Australia, Sweden, Italy and France banned their use some 20 years ago. The remaining countries of the EEC adopted the same policy in 1981 when regulations were introduced to ban the use of stilbenes and thyrostatic growth promoters. However, the continued use of implants of the 3 natural hormones, oestradiol, progesterone and testosterone and of 2 substances which do not occur naturally, trenbolone and zeranol, are permitted but there are stringent precautions relating to their use, particularly in respect of withdrawal well before slaughter.

The 1983 situation in Britain as summarised by the MLC is presented in Table 10.1.

TABLE 10.1. HORMONAL GROWTH PROMOTERS AVAILABLE IN BRITAIN

Hormone	Trade name	Category of stock	Availability	Withdrawal period
Androgens				
Trenbolone Acetate (TBA)	Finaplix	steers, cows heifers	prescription	60 days
Oestrogens				
Oestradiol–17B	Compudose	steers	prescription	0
Zeranol	Ralgro	steers, heifers, bulls	pharmaceutical supplier	70 days
Combinations				
TBA + oestradiol	Revalor	steers	not yet licensed	—
Testosterone + oestradiol	Implixa BF	female veal calves	prescription	90 days
Progesterone + oestradiol	Implixa BM	male veal calves	prescription	90 days

Zeranol is the only hormone listed on Table 10.1 that a farmer is able to buy from a retail pharmacy or an agricultural merchant. The combination of TBA and oestradiol was still under test and not yet available when the table was drawn up. All the other items on the list require a prescription from a veterinarian and all but one of the brands requires a withdrawal period varying from 60 to 90 days before slaughter to minimise residues in the carcass. The exception is Compudose 365 which works by releasing low levels of oestradiol from an inert silicon implant for about a year. The implants are not cheap by any means and prescriptions from veterinarians can cost up to £3 a tablet. Their greatest response is obtained with steers and liveweight gain can be increased by as much as 30 per cent as compared with 8 per cent in heifers and 6 per cent in bulls. This increase in liveweight gain, like that from the now banned stilbenes, comes from a higher food intake and a greater production on lean meat. Nevertheless the increase in liveweight gain as a result of implantation is proportionately greater than the increase of food intake so that there is also an improvement in food utilisation.

There are grounds for wondering whether there is not a measure of over reaction in the stipulated precautions that now cover the administration of hormones when this is at a very low level and takes the form of an aural implant so that the site of implantation is not a part of the carcass. Certainly in the instance of banned stilbenes there is a wealth of scientific literature that is conclusively against there being measurable residues in the carcasses of animals that have had aural implants and it is unlikely that the picture is any different with natural hormones.

Chapter 11

PRODUCTION SYSTEMS

(1) General Considerations

IT IS said, with truth, that there is only one profit in finishing a beef animal; cynics would claim, with the premium prices that have recently been paid for good rearing calves, that dairy farmers are taking all the profit from this class of beef. Certainly over the past twenty years there has been considerable optimism on the part of many producers, especially those who purchase stores for fattening, which has not always been justified. If the finishing of cattle was the principal undertaking on farms instead of the minor enterprise that it so often constitutes, more realistic prices would undoubtedly be paid for stores. Unquestionably the regularity of the annual increments in beef prices, when the guaranteed price system operated and the rises in beef prices which have similarly been a feature until comparatively recently of the Common Agricultural Policy, have engendered an attitude that soon there is bound to be something more in the pipeline for the fattener.

Ideally the calf producer should also be the finisher, because this would cut out transport costs and auctioneers' commissions, but there are comparatively few farmers who finish the animals they breed. One can understand the intensive dairy farmer, who has fully committed land and labour to milk production, wants to sell surplus calves at the earliest possible opportunity and also the upland suckler producer's need to sell calves at the weaner stage. But looking at the beef industry as a whole the plurality of ownership along the production chain constitutes an economic weakness, with too many middlemen getting their cut en route.

Some claim that the average Irish store has five owners before it reaches a British farm and though the situation is not as bad with home-bred stock, nevertheless there is a considerable proportion of finished cattle that have three or more owners before they are sold for slaughter. Even at this point of final marketing there is unnecessary expense because rather more than half the fat cattle in Britain are sold through auction marts, which not only entails the levy of the auctioneer's commission but also the costs of double transport between farm and mart, and mart and slaughterhouse. A feeder who can reduce the number of links between birthplace and slaughterhouse will be able to maximise the profits that can be taken from beef, even if he does no more than take the cuts that go to dealers, hauliers and auctioneers.

FINANCIAL CONSIDERATIONS

Apart from the physical pressures that necessitate selling calves from their farms of origin there are also financial considerations; such things as the amount of capital that can be tied up in a beef enterprise, the rate of turnover of this capital, and the expected profitability of the venture. The producer of barley beef based on the purchase of week-old calves is in a comparatively favourable position so far as turnover of capital is concerned because the whole cycle, under efficient management, is completed within the year. When a unit is a going concern with fairly regular intakes of calves there is a corresponding regularity in returns from finished animals so that there are not the troughs and peaks in capital requirements characteristic of the 18-month systems based on the purchase of autumn-born calves. Here the producer gets all his returns in the space of about three months covering the end of winter. Early in the New Year he will not only have a substantial investment in his nearly finished cattle, but also the investment in the next crop of calves which were purchased in the previous autumn. A sympathetic bank manager is an important asset in such situations.

Both barley beef and '18 months' producers can shorten the length of their investments by purchasing weaned calves at 3–6 months of age, according to the system, but this will increase the amount of capital invested per animal because the rearer's profit is added to the investment. In many respects, however, a division in the production cycle between the rearer of dairy-bred calves and the feeder has much to recommend it, provided there is contract selling between the two parties, unless the feeder himself has first-class rearing arrangements. Calf rearing has now become a very specialised job, requiring a very high level of expertise. The high level of mortalities and the loss of thrift in calves that characterised many rearing units when good Friesian and Friesian cross calves cost less than £10 can no longer be tolerated, and only the most competent are able to undertake this operation with any hope of financial success.

REARING CALVES FOR BEEF

Calf-rearing, based on the purchase of calves from many sources, usually through marts, cannot be other than a tricky business because the very young calf faces many disease risks. Scrupulous hygiene must be adopted and good nutritional regimes must be followed. It can, however, be a fairly profitable venture for the good stockman with limited capital since the investment per animal is relatively low and there is a quick turnover of money. Most rearers have a slack period in May–July when calves are in short supply, but it is not disadvantageous to have a period of up to two months when buildings are completely free of calves so that a thorough disinfection of premises and equipment can be undertaken.

The longest production cycle in beef production occurs where a farmer with a suckler herd finishes his own weaners. Starting from scratch with the purchase of bulling heifers he has no income for at least two years, and possibly longer, apart from any headage payments that may apply, for example that for breeding cows in less favoured areas, as defined by the EEC. The situation is usually more difficult for the owner of a spring-calving herd than for a farmer with an autumn-calving herd. Given the right choice of breeding stock, combined with good management the latter will be able to sell finished heifers at 13–14 months and steers at 14–16 months at a time of the year when prices for fat-stock are at a maximum. With a spring-calving herd there is inevitably a higher proportion of somewhat older cattle that will be sold off grass in the later summer–autumn period when prices are at their lowest level. There will however be some compensation in the shape of greater slaughter weights and lower feeding costs.

The overwintering of weaned calves as stores, either for re-sale in the spring or for finishing on the same farm, can, however, be a profitable proposition. Almost every autumn sale of suckler calves, and particularly those in marginal farming areas, has its quota of tail-enders, usually later born animals, which often sell at lower liveweight prices than the bigger calves that can be finished within the following six months. The opportunist buyer can sometimes get very good bargains, provided he can winter them cheaply and effectively. The arable farmer with otherwise unused buildings, bedding straw, seeds hay, some tail corn—and money in the bank from the sale of crops—can use these resources to advantage in this way. The sale of these stores usually on a seller's market at the end of the winter will release capital to finance the spring cropping programme.

GROSS MARGINS AS A MEASURE OF PROFITABILITY

At the outset a warning must be sounded on the dangers of an indiscriminate use of gross margins in comparing the merits of different production systems—a disease that has afflicted British agriculture for a number of years. The best example of the misuse of these statistics is to be found in cereal production, where favourable gross margins, particularly for barley relative to beef and sheep, encouraged many farmers, abetted by advisers who should have known better, to go into continuous cereal growing, often with disastrous results because of reduced yields caused by disease, competition from weeds such as couch grass and wild oats, and deterioration of soil structure.

The tragedy for many of these men is that the capital they formerly had invested in stock has been reinvested in machinery and equipment, which are a wasting asset, and now that they are being forced to go back to alternate husbandry they have to do this on borrowed money. The man who had £10,000 invested in an established suckler herd in 1962 who has maintained his herd at the same strength and with the same age distribution would have had 20 years later, as a consequence of inflation, an asset worth at least £80,000. Livestock, and especially breeding animals, are a hedge against inflation, but they are more than this on arable farms with heavy soils that are subject to structural damage under continuous cropping or on easier soils where there are no profitable alternative crops to break the sequence of cereals. This does not infer that one can afford to run cattle and sheep at a loss, merely to enhance cereal yields—to the contrary—but that the enterprises on such farms must be considered as an entity and not as isolated ventures.

Even gross margins comparing different sorts of beef production can be dangerous if expressed purely in terms of margins per animal or per acre, in the fashion adopted by the Meat and Livestock Commission in setting targets for farmers to achieve. Because of the differences in capital commitments that are characteristic of different systems, as well as rate of turnover of this capital, gross margins should be considered in terms of unit capital employment as well as on a per acre or per animal basis. Budgets or analyses of contrasting systems may show identical results on a per acre basis, but one alternative may show a substantial advantage over the other in respect of gross margin per thousand pounds of invested capital. Again, one must take into account the effect that one system may have, as compared with another, on fixed costs, which are by no means fixed in practice. One system may put greater demands on available labour than another or require more buildings, which become a fixed cost charge against the enterprise, and this must also be taken into account in decision making.

Gross margins are probably most safely interpreted when considering modifications of an existing system rather than a change to an alternative system, for instance the consequences of raising stocking intensity in a suckler herd. But an improvement in gross margins from £220 to £300 per ha can be counter-profitable if the bank charges on additional cows and the buildings to accommodate them swallow the whole of the projected increase of £80 and a bit more besides. Farming has enough intrinsic challenges without accepting that of working for the bank. Unquestionably gross margin estimates have a value in farm planning but their message must be taken with the appropriate grains of salt.

RELATIVE PROFITABILITY OF DIFFERENT SYSTEMS

One must accept that some systems of production are more suited than others to a given farming situation because of the physical situation of the farm and the management expertise of the farmer. For instance, an upland or hill farmer has no real choice apart from the production of suckled calves; the most he can hope to do beyond this is to over-winter a proportion of his younger weaners for sale in the spring. Apart from such considerations as food supplies, the existing subsidy system is all in favour of breeding rather than the growing on of stores on such farms. On lower land the farmer has a much greater choice—the production of barley beef, intensive grass-dairy beef, the finishing of purchased suckled calves, the buying in of older stores for finishing on grass or in yards, and even the maintenance of a suckler herd combined with finishing of the calves.

In the long run one cannot say for certain that any one of these alternatives is a better bet than another, because immediately one approach is shown to be relatively more profitable than another there will be a swing in this direction that will push up costs. Probably the best example of rapid erosion of profit margins is provided by barley beef which within a short time of the publication of Dr Preston's initial results became the most profitable form of beef production at that time with a net return in a 12-month period of approximately 30 per cent on capital invested in the enterprise. Naturally there was a very quick reaction and as more and more farmers adopted the system the prices of calves and barley progressively hardened to the point where a producer was doing very well if he cleared 15 per cent on his investment. The expansion of barley beef production was not the only factor in pushing up the price of Friesian and Friesian-cross calves. From about 1955 onwards there was a growing realisation that these were also valuable animals for more conventional feeding systems.

At first the emphasis with pure Friesians tended to be placed on the slaughter of heavy cattle in the weight range of 600–50 kg liveweight at 2½–3½ years but from the late fifties it moved with the development of the 18-month system to a slaughter weight of about 500 kg which produces a carcass that is about 50 kg heavier than the typical barley beef carcass. This is an important consideration nationally for apart from a better spread of the initial cost of the calf it gives a fuller realisation of the calf's beef potential.

The day of the really heavy carcass seems to have passed except perhaps for the kosher trade. The emphasis in the retail meat trade is now firmly placed on carcasses from younger animals with mild-flavoured cuts of tender meat. Apart from this it does not pay farmers to hold animals for long periods with little or no return on the investment and so we no longer see the once common situation in arable districts where yarded cattle were fed cheap rations, often consisting of by-products to do little better than maintain their weight over the winter. Their function primarily was not one of providing direct profit but of treading straw into farmyard manure. With the availability of high quality conserved forage (mainly grass silage and possibly in some southern districts maize silage), arable farmers are able to produce farmyard manure where this is deemed to have a special value for root crops and also to make a modest profit from a beef enterprise.

PEDIGREE BREEDING AND IMPORTED BREEDS

The breeding of pedigree livestock understandably has a great attraction for many farmers and this does not depend entirely on the financial rewards that the successful breeder can earn when he reaches the top of his particular breed pyramid, for there is an enormous satisfaction to be derived from breeding animals that are considered to be first class specimens of a breed. It is also significant that most successful breeders have an intimate knowledge of their stock and their ancestry and progeny for this is not only indicative of their dedication to their task but also a reflection of the depth of their interest in their work. Top breeders not only have status, for invariably they are also very happy men and it is not surprising that young men, still on the lower rungs of a farming ladder, often nurse ambitions to join the ranks of the top men in the breed of their choice.

Entry into this circle of elite breeders can be a very expensive operation. It is one thing for the successful industrialist or pop singer to buy an estate, hire a knowledgeable manager and spend large sums on purchasing foundation stock because in such instances money is no object but it is a very different matter for the working farmer who hopes to break into the pedigree breeding world. Good pedigree stock carry a substantial price premium over commercial stock but investment in them is only viable economically when the new breeder in his turn is able to get premium prices for the stock he sells.

The situation for the new entrant into pedigree breeding of beef cattle is undoubtedly easier now than it was 30 years ago. At that time there was a large measure of localisation of the top herds of a given breed. For instance most of the leading breeders of both Aberdeen Angus and Beef Shorthorns were situated on the east side of Scotland in Perthshire and further north into Aberdeenshire, while the majority of the top Hereford breeders were to be found within a comparatively short distance of Hereford. Many of the top herds in all the principal breeds were long established and often had passed from father to son. Reputations had been created by successes in both show and sale rings. The breeder of the top Aberdeen Angus bull at the Perth Sale, for instance, brought further recognition for the herd, not only in Britain but in other beef producing countries as well. The top breeders tended to form a tight circle that practised a judicious purchase of bulls of each other's breeding. It has been alleged that such purchases were sometimes finalised before the bull in question went under the hammer at the leading breed sale and that the enhanced price there was no more than a way of drawing attention to the two herds that were involved in the transaction.

The introduction of the Continental beef breeds and the development of more objective measurements of performance in beef cattle totally altered the situation and have made it easier for new entrants into pedigree breeding to create reputations for their breed, always providing they have the necessary capital to purchase adequate foundation stock and they are also first-class stockmen. Performance testing and herd recording have not completely disposed of the hierarchal structure within the older breeds but they have certainly made it easier for the young progressive to make a name for himself.

The introduction of exotic breeds has resulted in a lot of new recruits to pedigree breeding of beef cattle. Some of the earliest entrants who staked their claim before the first importations of Charolais females arrived in Britain in 1966 were looking for a quick profit. At that time there was a strong unsatisfied demand for Charolais cattle from breeders in the United States who were unable to buy directly from France, and Britain came to be regarded by some as a transit quarantine station and stock was often exported at very advantageous prices. On top of this there was also a feeling with many who were early on the Charolais band-wagon that the example of high profits made by the first breeders of the Swedish Landrace pigs in Britain was to be repeated and as a consequence many of the first people to own Charolais in Britain were speculators rather than breeders.

Fortunately for the breed this situation has changed completely as the Charolais came to be recognised for what it is, a very good top crossing breed which, mated to a wide variety of suckler cows, will produce cross-breds with excellent growth performance and good slaughter attributes. Many of the established breeders are, in addition, concerned with the production of suckled calves and the acquisition of Charolais stock has been essentially an expansion of their cattle business. Their plans in this connection have been greatly helped by the availability of semen from bulls that had been vetted by the Milk Marketing Boards and other AI authorities in respect of ease of calving as well as being selected as good specimens of the breed. In consequence at a time when Charolais were expensive to buy a potential breeder was able to concentrate on female purchases until such time as the purchase of a bull was justified. There is no area in Britain that can be described as the centre of gravity of the breed for there is now a nation-wide spread of breeders. Naturally they are to be found in their greatest numbers where the production of sucklers is important for instance on either side of the Scottish border.

No sooner had the Charolais made its mark in Britain than barriers against the importation of other breeds principally from Western Europe were lifted. Here is a complete list of new arrivals with the year in which they were first imported given in parentheses: Blonde d'Aquitaine (1972), Chianina (1973), Gelbvieh (1973), Limousin (1971), Marchigiana (1974), Maine Anjou (1972), Murray Grey (1973), Romagnola (1974), Simmental (1970), Normandy (1975), Rotbunte (1976) and Pinzgauer (1976). The list of imported breeds can be increased to 14 if one includes the MRI (Meuse-Rhine Ijssel), a red and white dual-purpose breed from the Netherlands which was imported several years ago to become part of a breed synthesis programme with the Dairy Shorthorn in an effort to save that breed from possible extinction. It is unlikely that many of the breeds in the above list, apart from the Simmental, Limousin, Blonde d'Aquitaine and possibly the Murray Grey, will find a permanent place in Britain alongside the 14 existing beef breeds of purely British origin. It is interesting that of the above Continental breeds there were no less than 8 where there were exports to other countries, mainly Canada and USA within 3 years of their original importation. Clearly there was a speculative aspect in these importations, with Britain being used as a transit area in satisfying the North American's appetite for exotic breeds. One cannot but wonder how many will be represented in Britain in 1990.

Chapter 12

PRODUCTION SYSTEMS

(2) Suckler Calves

SUCKLED CALF production is essentially a low output system of land use which, on the whole, is best suited to marginal land. Table 12.1 which gives gross margins for lowland and upland herds over the period from 1978 to 1982 illustrates this point. Indeed when a comparison is made with the gross margins per ha that typical lowland farmers were achieving over the corresponding period, up to 3 times greater than those for suckled herds, it is questionable whether such an enterprise has a place on lowland farms except perhaps where it can be fitted in as a convenient sideline of an intensive and highly profitable main line of production. For a variety of reasons many lowland farmers with grassland at their disposal are not interested in dairy production and this may account for the choice of a suckler enterprise but there are other forms of beef production potentially more profitable, for instance 18-month grass–cereal beef. Regularly the top operators in this branch of beef production achieve gross margins of the order of £800 per ha.

TABLE 12.1. GROSS MARGINS FOR LOWLAND AND UPLAND SUCKLER HERDS 1978–82 (MLC)

| | Gross Margin (£) | | | | | |
| | Lowland | | | Upland | | |
	per cow	per ha	adj per cow*	per cow	per ha	adj per cow*
1978	109	205	180	133	202	219
1979	113	208	161	140	218	199
1980	118	207	142	154	223	186
1981	132	245	143	186	227	202
1982	145	278	145	213	270	213

*Adjusted for inflation using General Retail Price Index

There would be less justification for this rather categoric view that beef breeding cows are better suited to the marginal land if it were possible to establish that attention paid to a few key factors should create a spectacular increase in returns. This does not appear to be the case judging by the data given in Table 12.2,

TABLE 12.2. PERFORMANCE OF 81 LOWLAND HERDS (1982, MLC)

Financial results £ per cow mated	Average	Top-third
Output		
Calf sales	260	287
Cow subsidy	11	11
Herd replacement	−22	−21
Output	249	277
Variable costs		
Cow concentrates	20	13
Calf concentrates	11	12
Forage	40	38
Other feed	13	10
Vet & medicines	9	8
Bedding	7	7
Other costs	4	4
Total variable costs	104	92
Gross margin/cow	145	185
Gross margin/ha	278	394
Physical results		
Calves purchased (%)	5	2
Calves reared (%)	92	93
Average weight of weaner (kg)	270	286
Stocking rate (cows/ha)	1·9	2·1

derived from a 1982 survey of lowland suckler herds. The top-third have an average gross margin of £394 per ha which is slightly more than 40 per cent greater than the average figure for the whole sample of £278 per ha. However when one examines the components of the input and output data there is no one factor that appears to be of outstanding importance. Instead there is the cumulative effect of such factors as a higher stocking intensity (2·1 v 1·9 cows/ha), higher selling price (£287 v £260) and lower concentrate costs (£13 v £20/cow). A very similar picture emerges from a survey of 95 upland herds where the average gross margin per ha is £270 as compared with £380/ha for the top-third. The only certain conclusion to be derived from these figures is that relative success, as judged by the level of gross margins, either per cow or per ha, depends on a complex of factors which individually are of relatively little importance but collectively have a considerable influence on margins but not to the point where suckler production can be regarded as a logical alternative to dairying or 18-month beef on lowland farms.

It is important to remember that gross margins, which are the difference between the value of the product and the variable costs that are involved, are only part of the story because the ultimate factor in determining profitability are the fixed costs or overheads. These include the permanent labour force, the interest on and the depreciation of the investment in land and buildings, the rental value of the land, whether this takes the form of actual rent or mortgage payments and bank charges. Even at high levels of output a suckler herd can only be profitable if there is a low cost structure that includes fixed as well as variable costs.

CHOICE OF BREEDING COWS

Human nature being what it is, with personal likes and dislikes and traditional loyalties, it is inevitable that there should be a very wide spectrum of choice for both suckler dams and sire breeds. A Welsh hill farmer will need a lot of persuading if he is to change from the native breed that has satisfactorily served him and his father before him over the years. He has considerable justification for his choice for the Welsh Black is a relatively hardy breed, a good milker and given the right sort of management it will rear an excellent calf. Correspondingly a Scottish farmer who has had good results from Blue Grey cows over many years is disinclined to switch to another breed or cross, particularly if he is on an upland farm, where hardiness and hybrid vigour have a special significance. The data in Table 12.3 comparing the performance of Blue Grey cows with Hereford × Friesians when mated to the Charolais justify his preference.

TABLE 12.3. COMPARATIVE PERFORMANCE OF HEREFORD × FRIESIAN AND BLUE GREY SUCKLER COWS MATED TO CHAROLAIS BULLS (MLC)

	Hereford × Friesian	Blue Grey
Cow weight (kg)	500	450
Assisted calvings (%)	10·1	9·0
Live calves (per 100 cows)	94·9	95·6
Calving intervals (days)	378	369
Calf weaning weight (kg)	294	278
Calf weight/cow bulled (kg)	269	265
Calf weight/50 kg cow wt (kg)	26·9	29·3

The Blue Grey has a slight advantage over the Hereford × Friesian in every respect except weaning weight of calves, where there is an advantage of 16 kg in favour of the weaners out of Hereford × Friesian dams. This disappears, however, when calf weight is related to weight of dam—an important consideration especially under upland conditions because the lower liveweight of the Blue Grey means that it has a lower maintenance requirement than the larger Hereford × Friesian cow. The more regular breeding pattern of the Blue Grey is another important attribute and the cross is also characterised by a long working life with an average of 9 lactations. Unfortunately there is only a very limited supply of Blue Grey heifers and this means they are expensive to buy.

At one time there was a regular supply of Irish-bred heifers, usually by Aberdeen Angus out of Dairy Shorthorn cows, and these found great favour particularly in the Border counties. A prized virtue was their tractability, mainly attributable to their being bucket-fed and used to handling during the course of rearing. This is an important management consideration especially when dealing with difficult calvings or setting on calves to replace neo-natal casualties. With the virtual disappearance of the Dairy Shorthorn in Ireland and its replacement there by the Friesian the traditional Irish heifer is now little more than a memory. Its place has been largely taken by the Hereford × Friesian or less commonly, by the Angus × Friesian. These are both capable also of rearing good calves and since they are mainly hand reared they are very tractable. Both crosses are criticised on the grounds that they can have too much milk for their calves and also they are said to be rather 'thin in the skin' as compared with the Irish cows they replaced.

Apart from the Welsh Black, several of the localised British breeds are also used as purebreds for suckled calf production, in particular the Sussex, Devon and the Lincoln Red. Possibly their role in this respect owes more to tradition than to any great virtues as dams of suckled cows.

CHOICE OF SIRE BREED

There is no lack of choice among sire breeds for use on suckled cows. The principal considerations here are the growth rate of calves, their conformation and relative ease of calving. On the first point, reference back to Table 3.5 will recall that 2 Continental breeds, the Charolais and the Simmental are matched by only one British breed, the South Devon, in respect of weaning weights under lowland, upland and hill farming conditions. Unfortunately Table 3.6 also reveals that these 2 breeds have higher rates of calf mortality and a high incidence of assisted calvings as compared with Hereford or Aberdeen Angus matings. Simmental, Charolais and Limousin crosses all have high carcass conformation scores while the South Devon compares badly with all other breeds in this respect.

The Limousin is outstanding as a butcher's beast for it has a high killing-out percentage and a high proportion of the more expensive cuts, see Table 7.1. Unquestionably if it was a colour-marking breed, like the Hereford or Charolais, it would occupy a more prominent position than it presently has. The big two, however, are unquestionably the Charolais and to a lesser extent the Simmental which are continuing to make steady progress at the expense of the British breeds. Breeders in both instances are making ease of calving an important selection criterion and farmers have learned the importance of using a bull of a safer breed such as the Aberdeen Angus or the Hereford on maiden heifers, and in the case of older cows mated to the Charolais while keeping them in a relatively lean condition coming up to calving to reduce birth weights.

AGE OF FIRST CALVINGS

It is a common practice in suckler herds to calve heifers for the first time at the 3-year-old stage but we believe, in the light of experience in the dairy industry, that this is wasteful of both time and capital investment. Milk Marketing Board surveys reveal no adverse effect on life-time performances as a result of calving dairy heifers at 2 years as opposed to 3 years of age and it is reasonable to expect the same result for beef herds, providing heifers have been well reared because sexual maturity is influenced by level of nutrition as well as by age. Where, of necessity, heifers have to be reared under hard conditions, for instance Blue Greys on hill farms, calving at the 3-year-old stage may be the only feasible decision but this need not be the case with beef-dairy breed crosses reared under easier conditions.

Heifers mated at 15–18 months should have preferential treatment until at least 3 years of age. It is advisable to mate them to one of the low birth weight breeds, eg. Aberdeen Angus, Hereford or Sussex, and to keep them on no more than a moderate level of feeding as they approach parturition to minimise calving difficulties. Once they have calved, however, their plane of nutrition should be stepped up to ensure that they put on weight in advance of mating. If they are in low condition when their calves are coming up to 8 months, immediate weaning is recommended to give them a 4-month dry period so that they come to their next calving in reasonable condition.

Experience with dairy heifers of several breeds in many parts of the world has shown that subsequent breeding performance is impaired if there is delayed mating of heifers especially if they have been reared on a high nutritive plane. This is not generally a problem with mating at 15–18 months of age but if mating does not take place until 27–30 months of age it is wise to adopt a moderate feeding regime.

SELECTION OF BULLS

Apart from exercising a breed option it can be false economy to purchase a bull within that breed just because it is cheap. A bull can leave upwards of 200 calves with natural service and the difference in financial terms between the progenies of good and bad bulls at weaning could be more than £2,000. Fortunately with the development of central performance testing and of on-farm recording in bull breeding herds prospective buyers now have the opportunity of basing their selections on performance data. Weight-for-age and conformation, however, are not the only considerations, especially with the larger Continental breeds. An endeavour should be made to ascertain whether the bull's sire has a good record in respect of ease of calving. Particular attention should be paid to the quality of the legs and feet and especially the hind legs because physical disabilities of this nature can impair mating performance.

TIMING OF CALVING

There are, in theory at least, two principal calving seasons, the autumn and the spring. There are many instances, however, where herds, initially intended to be autumn calving, have a high proportion of winter calvings because of delayed conceptions which are discussed in Chapter 6 and are often attributable to weight losses during the intended mating period. The problem is less serious in spring-calving herds, always providing bulls are fertile, because mating takes place at a stage of the year when pasture is usually plentiful and of good quality nutritionally.

Because grazing is the cheapest source of food for suckler production it would seem logical to calve the herd about 6–8 weeks in advance of active pasture growth for this, with silage as the main feeding stuff until the availability of adequate pasture, would give the best synchronisation between herd appetite and pasture growth. There are other considerations however, in particular the value of the individual calves in the autumn sales. Big calves, the consequence of greater age as well as breeding and management, sell at a premium because they can be finished in a matter of a few months, with a quick turnover of the investment, to meet a market at a point when prices are normally at their highest. This is illustrated by the figures given in Table 12.4.

Though it is unwise to draw anything but tentative conclusions from the data, particularly that for summer and winter calving,

TABLE 12.4. 1982 GROSS MARGINS AND PRICES FOR SUCKLED CALF PRODUCTION ON
LOWLAND AND UPLAND FARMS ACCORDING TO SEASON OF CALVING (£)

	Lowland Farms			Upland Farms		
Calving season	No of herds	Calf price	Gross margins* per cow per ha	No of herds	Calf price	Gross margins* per cow per ha
Spring	33	229	138 276	35	253	193 284
Summer	13	268	134 234	10	289	213 264
Autumn	20	310	163 286	31	347	242 281
Winter	13	281	155 335	8	267	192 261

*The subsidy on upland farms was £52 per cow as opposed to £11 on lowland farms

because of the relatively few herds that are involved, nevertheless
under both upland and lowland conditions the gross margin per
cow is appreciably higher for autumn calving than for spring
calving (£163 v £138 and £242 v £193 respectively). However
differences in gross margin per ha are of a rather different order
with £286 for autumn-calving lowland herds as against £276 for
spring-calving herds and £281 and £284 respectively for upland
herds. The explanation for this is the higher stocking rates with
spring calving (2·0 cows/ha) as compared with autumn calving (1·8
cows/ha) on lowland farms. The corresponding figures for upland
herds are 1·5 cows/ha and 1·2 cows/ha respectively for spring- and
autumn-calving herds. Another management factor of considerable significance, particularly with upland farms, is that autumn-
calving cows require 1·5 tonnes more silage per head as well as
considerably more concentrates than spring-calving cows. Certainly in terms of land utilisation spring calving is preferable to
autumn calving.

BLOCK CALVING

Disregarding season of calving, the ideal is to have a calving
pattern where all the calves are born within a few weeks of one
another. Apart from this resulting in a more even bunch of
weaners at point of sale, management is simplified in that the herd,
with its varying nutritional requirements over the year, can be
treated as a unit.

The principal factors in achieving this ideal are: (a) making
certain that a bull is sound on his feet, is fertile and has the
necessary libido; (b) ensuring that the herd is in reasonable
condition at mating, a point that will be discussed in greater detail
in the following section; (c) restriction of the mating period to no
more than 10 weeks once out-of-phase females have been culled
from the herd. Generally it is advisable to calve heifers entering

the herd a few weeks in advance of the main body to give a little more latitude in respect of delayed conception and to give them preferential treatment during their first year to allow for growth as well as lactation.

HERD PERFORMANCE AND LEVEL OF NUTRITION

In Britain the normal aim of a dairy farmer is the maximising of milk output by high level feeding, especially at certain critical stages of the production year, for instance steaming-up immediately prior to calving and again immediately after calving as cows are coming up to their peak yields. This approach has been economically feasible because of favourable relationships between milk and concentrate prices but this situation may alter with the imposition of production quotas for these may necessitate a reappraisal of input–output relationships in dairying and a greater reliance on cheap bulk foods. This, however, will be only a mild form of the position that the suckler calf producer has always had to face. Indeed there are many parts of the world where beef breeding herds have to make do without any form of feeding, with the possible exception of mineral mixtures, that supplement intakes from grazing or browsing. Conditions are not as demanding as that in Britain but nevertheless grass as grazing or conserved as hay or silage must constitute the principal source of nutriment for suckler herds. As far as conserved grass is concerned, there need be less emphasis on quality than there is in dairying and this makes possible a greater emphasis on quantity especially where later winter–early spring calving is practised.

There is a limited role for concentrates: typically spring-calving cows generally have an annual allowance of about 100 kg which is about half the level of concentrate feeding usually practised with autumn calving. Many suckler producers particularly those practising autumn calving prefer to feed a high proportion of the concentrate allowance directly to the calves as a creep ration rather than to the dams. There is good sense in this for the double processing of concentrates into milk and then by the calf into flesh is not an efficient operation. Often concentrate feeding for calves is reserved for the tail-end of lactation and immediately after weaning to ensure that there is no post-weaning check or loss of bloom that will impair their sale value. This is particularly appropriate with autumn-calving herds for here weaning normally takes place some weeks in advance of the autumn suckler sales. Many buyers have a preference for calves that have their weaning traumas behind them and are conditioned in part to the type of feeding that will follow under their new ownership, over those that are freshly weaned and often subject to transit fever.

Taking a leaf out of the book of dairy husbandry advisers, MLC officers now use a system of condition scoring as a basis for their advisory work. * Body condition is graded on a 6-point scale from 0 to 5 along the following lines:

0. A condition of severe emaciation with the spine very prominent and no detectable fat cover over transverse processes in the loin region which are sharp to touch.

1. Spine is prominent but the transverse processes are less sharp.

2. Transverse processes are still distinct but are rounded from a covering of fat.

3. The transverse processes are only detectable with firm pressure.

4. Transverse processes cannot be felt even with firm pressure and the cow is in prime slaughter condition.

5. The cow is grossly overfat.

Scoring, in the hands of an experienced operator, is a useful method of quantifying body condition and providing targets at various important stages of the production cycle. The MLC postulates a target score of 3 for autumn-calving cows, that is to say a forward store condition, on the grounds that there is some scope for cows to milk condition off their back and thereby permit some economies in feeding. Normally there should be no great difficulty in getting this level of condition with an October calver that has been dry for at least two months because cows after weaning usually put on a lot of flesh and sometimes the problem is that they are in such high condition that there may be difficult calvings, particularly with heavy breed matings. The suggested target figure is 2·5, ie. a good store condition, for mainly spring-calving cows and this should be feasible with an intake up to 30 kg of reasonable quality silage (DM 25%, ME 10) plus a picking of good barley straw in the 3 months prior to calving.

The critical stage of the production year is at mating and here the suggested condition target for autumn-calving cows is 2·5. Probably the first few weeks of lactation can be based on pasture but as this starts to fail silage feeding should be increased progressively up to 30 kg per day. Then if the cows are thought to have insufficient flesh 2 months after calving a supplement of about 2 kg of mineralised rolled barley may be advisable. Once

* See MLC publication *Blueprints for Suckler Beef.*

mating is successfully accomplished it is permissible for cows to lose some condition to give turnout rating of 2. If the available grazing is adequate for the needs of the calves their dams should have no difficulty achieving a target score of at least 2·5 at weaning.

Similarly with spring-calving cows, if mating commences in mid-May for calving in February–March there should be no problem in lowland herds of achieving a condition score of at least 2 but preferably a little higher. With earlier calving (January–February), which many producers prefer to get well-grown weaners for the autumn sales, it may be necessary to feed at least 30 kg of silage plus a kg of mineralised barley in order to reach target condition. This will also hold with hill herds even with March–April calving for there is seldom sufficient grazing even on in-bye fields to satisfy the requirements of the herd until early June.

HOUSING

There is an old saying that cattle have 5 mouths in the winter and this is particularly true with cattle on strong land for poaching not only fouls any existing grazing but also destroys much of the potential growth of the following spring. This situation is of less concern where rough grazings are being progressively improved by surface measures for hoof cultivation, combined with the application of lime and phosphate can be a cheap and effective method of breaking a mat of poor herbage over accumulated vegetable residues that is so often characteristic of moorland pastures. It is a different matter when pastures have been improved and stocking intensities have been raised for then poaching becomes counter-productive. In the absence of convenient sacrifice areas such as an old quarry field in permanent pasture, with rocky outcrops, or links land where stock can shelter among sand dunes in bad weather, consideration must be given to some form of winter accommodation for the sucklers. It must not be expensive housing, such as a wide-span canopy building, unless there are other uses for it apart from wintering sucklers, such as the storage of grain that will be sold before the onset of winter, because the margins in single suckling are insufficient to carry high interest and depreciation charges.

Sometimes a farm is in the fortunate position of having existing buildings that can be adapted to service the need, for instance kennels with open yards. These however require a supply of cheap straw for bedding so they are of little use on farms where straw is such an expensive item that it is primarily used for fodder. Cow kennels with cubicles probably constitute the cheapest arrangement under these circumstances. When in-wintered cows have calves at foot it is advisable to have a bedded creep area so calves can be given supplementary food as well as have a separate resting area. An area of approximately 25 sq metres will accommodate 20 calves. Loose housed cows require 6–7 sq metres per head (though not all this needs to be covered) and at least 75 cms of trough space where there is rationed feeding. Obviously with a spring-calving herd winter accommodation need not be so elaborate as that required for an autumn-calving herd. Indeed in Eire there are many farmers who successfully winter cattle in topless cubicles and these could be feasible on sheltered sites in Britain. However with uncovered accommodation there is the inevitable problem of slurry diluted by rain water for this, in a society now very concerned with the avoidance of river pollution, cannot be allowed to go directly into streams.

DISEASE AND RELATED HAZARDS

Calves that are housed in their first month of life are vulnerable to infections especially those causing scouring and, in badly ventilated buildings, pneumonia. An additional reason for farmers favouring autumn calving is that weather and soil conditions are usually such that calving can take place out-of-doors and it will not normally be necessary to bring calves indoors before they are 4–5 weeks old when scouring is much less of a hazard. Pneumonia, however, will continue to be a potential danger even with the older calves, particularly in the November–December period which so often is characterised by high humidity.

Now that brucellosis is substantially under control, hypo-magnesaemia (tetany or grass staggers) is potentially the biggest hazard in a suckler herd. It can occur at any time of the year but it is most prevalent in the late winter when there is no grass available, in the spring following a rapid onset of grass growth or, with freshly-calved cows in the autumn. A disconcerting feature of the disease is that the average loss attributable to it is not high and one can go for several years with complete freedom from trouble and then it strikes with the loss of several cows within a matter of days. Older, deep milking cows are the most vulnerable. While it is most uncommon for the acute form of the disease to affect heifers, they may be subject to a milder form that is characterised by nervous reactions. The incidence of hypomagnesaemia is aggravated by stress, for example cold, windy weather after calving can precipitate trouble.

The condition occurs when there is a prolonged period where a cow's intake of magnesium is less than her output, a situation that is likely to occur when freshly-calved cows have been on a diet of indifferent hay especially if it has been subject to weathering with loss of soluble nutrients. It can also occur with a grass dominant pasture that is making rapid growth on land that is either naturally rich in potash or has had a recent application of potash. Under these conditions, particularly if a nitrogenous fertiliser is also applied to boost growth, there can be a luxury uptake of potash in the sward at the expense of magnesium and other bases. Once clover makes an appreciable contribution to grazing intake, as it usually does in the middle of the growing season, the danger subsides because clover has a high magnesium content as compared with grass. As a form of insurance, where potash is a limiting factor in grass growth, it is safer to apply it during the summer rather than the early spring if the pasture is to be grazed by cows in milk.

The feeding of a magnesium-enriched supplement to animals at risk is a sensible precaution. A daily intake of 60 gm of calcined magnesite will suffice but since it is unpalatable it has to be incorporated in supplementary food, for instance concentrates fed at the rate of 1·5 kg per day. This may be justified in the winter before grass growth starts but there are economic limits to the amount of concentrates that can be fed to suckler cows. Another measure is the feeding of proprietary high magnesium pencils but they are expensive and can only be justified where there is a risk of a clinical manifestation of disease. It will be cheaper and possibly just as efficacious to provide a mineral supplement containing magnesite. An alternative where silage is being fed is to sprinkle magnesite on the fodder. Alternatively it can be included in the herbage when it is being ensiled. An addition of 2 kg of calcined magnesite to each tonne of green material as it goes into the clamp will not have a deleterious effect on fermentation and it will ensure that the level of magnesium in the blood does not fall below danger level.

PRODUCTION TARGETS

The MLC has been able to establish production targets for single suckling herds over a range of farming conditions based on farm records collected over a number of years. These are given in Table 12.5 in relation to 2 sire breeds, the Charolais and the Hereford and the two principal calving seasons, autumn and spring.

Except for stocking rate, the same targets are suggested for both lowland and upland herds and this is reasonable because with good management there is no basic reason why production on marginal farms should differ materially from those on better land. It is a rather different situation on hill farms for here the harsher conditions are bound to affect conception rates, calf survival, daily gains and weight at point of sale. In every instance the weight-at-sale target is higher for autumn-calving herds than it is for spring herds but this is mainly attributable to autumn-born calves being older at point of sale (usually in October) than spring-born calves. This in its turn requires greater feed inputs in the shape of concentrates, and conserved grass, as well as some additional grazing. On this account one cannot but question the MLC setting targets for autumn-calving herds on hill farms where the growth of pasture is limited to no more than 5 months of the year while invariably there is only a limited area of in-bye grassland that can be used for conservation. Possibly the most sensible approach under typical hill farming conditions is April calving and October weaning with immediate disposal of calves, so that the cows are under any degree of physiological pressure for no more than the 7 months of the year that span the growing season. Lower realisation for the individual calves could be more than offset by an increase in the size of the herd, made possible on this account and by a reduction in feed costs.

TABLE 12.5. MLC PERFORMANCE TARGETS FOR SUCKLER HERDS

Sire	Charolais		Hereford	
	Autumn	Spring	Autumn	Spring
Lowland herds				
Calving period (days)	90	85	85	80
Calves weaned (%)	92	92	95	95
Calf daily gain (kg)	1·0	1·2	0·9	1·0
Sale weight (kg)	350	300	300	250
Stocking rate (cows/ha)	1·9	2·3	2·0	2·4
Upland herds				
Calving period (days)	90	85	90	85
Calves weaned (%)	92	92	95	95
Calf daily gain (kg)	1·0	1·2	0·9	1·0
Sale weight (kg)	350	300	300	250
Stocking rate (cows/ha)	1·5	1·9	1·6	2·0
Hill herds				
Calving period (days)	100	90	90	85
Calves weaned (%)	85	85	90	90
Calf daily gain (kg)	0·9	1·0	0·8	0·9
Sale weight (kg)	310	265	260	220

Chapter 13

PRODUCTION SYSTEMS

(3) Grass-fed Beef from Dairy-bred Cattle

THE EARLIEST feeders of Friesians found that, on an all-grass diet in the summer and store feeding in the winter, it took 2½ to 3 years before they reached slaughter condition, generally in their last summer at grass. Admittedly they reached heavy weights, with 640–60 kg not uncommon, but the turnover was slow and they did not sell well in competition with beef crosses marketed at the same time, unless there happened to be a shortage of beef. Better results were obtained with Hereford crosses which would finish on grass or bulk foods at 18–24 months of age, admittedly at lower weights, but this only pointed to the necessity of better feeding regimes for the increasing numbers of straight Friesian stores that came in the wake of the early weaning system of calf rearing that was developed by Preston in 1955.

DEVELOPMENT OF THE 18-MONTH SYSTEM

We more or less stumbled on what is now known as the '18-month system' of grass-cereal beef production in 1956–58, in the course of work at Cockle Park relating to relationships between dairy and beef attributes in Friesians that has already been referred to in Chapter 5. We wanted information on the autumn-born calves as quickly as possible but in common with other feeders, we found that only a small proportion of each progeny group finished on pasture at the two-year-old stage and this necessitated winter finishing. Until then the diet for the second winter was exclusively good quality silage fed to appetite along with one feed of hay in the evening. The best we could get was 600–700 g of gain per day and though the animals grew well there was insufficient finish to consider slaughter at the end of the winter. Most of them continued to grow rather than fatten in their second summer and so the decision was taken to feed the 1956 intake on a silage plus barley diet over the 12–18 months stage and this worked admirably.

Meanwhile other centres and farmers too, notably Fenwick

Jackson then at Kirkharle in Northumberland, and later at Shoreham near Cornhill-on-Tweed, were experimenting with cereal-silage diets for this class of animal, but the real boost for the system came after Preston's advocacy of barley beef. He pointed out somewhat trenchantly that traditional grass feeding of cattle produced no more than 440 kg of gain per hectare whereas a hectare in a good crop of barley, with a quite small addition of protein, mineral and vitamin supplements, produced over 740 kg of gain.

His words stung workers at the Grassland Research Institute at Hurley into action, and in 1962 they started an intensive grass beef unit which was stocked with autumn-born Hereford × Friesian steers grazed rotationally on pastures liberally treated with nitrogen. Stocking rate was approximately 0·28 of a hectare per beast and surplus grass was saved as silage. The original aim was to finish the cattle at 12–15 months at a weight range of 390–430 kg, if possible off grass with silage coming in when grazing started to run short.

The project was only partially successful so far as the original objectives went, but output per acre more than matched Preston's figures. In common with previous results it was found that silage alone was not a good enough finishing diet even with Hereford crosses, and there was the added drawback that cattle sold in the late autumn met a slack market. The decision was taken to feed some barley over the final stages and so the end result was a characteristically British compromise of grass and barley.

This work did much to formalise the system though there were already many who had arrived at the same conclusion as the Hurley workers. The system spread rapidly as barley beef became less profitable primarily because it did not spread the initial cost of the calf sufficiently. The situation was and still is that we have not enough rearing calves in Britain to be able to afford to slaughter many Friesian and Friesian-cross cattle at 400 kg. Admittedly the 18-month method slows down the turnover of capital and gives less efficient food utilisation, but economically it is the better proposition not only because of the heavier slaughter weights, but also because of the substitution of a cheaper food in the form of grass as grazing, silage or hay for the more expensive cereal plus supplement mix.

INFLUENCE OF BRA AND MLC

Early on in its activities the Beef Recording Association collected and analysed farm data relating to the system and they were able to establish production targets and the important husbandry and management practices in the achievement of these targets which are those given in Table 13.1.

These targets envisage a use of 200–50 kg of elemental nitrogen per acre, applied at intervals over the season.

TABLE 13.1. MLC TARGETS FOR SEMI-INTENSIVE GRASS-CEREAL BEEF—AUTUMN-BORN STEERS

	Friesian	Hereford × Friesian
Daily gain (kg)		
Rearing to 12 weeks	0·8	0·8
12 weeks to turnout	0·7	0·7
At grazing	0·8	0·9
Finishing winter	0·9	0·8
Overall	0·8	0·8
Stocking rate (cattle/ha)		
Grazing		
To mid-season	10·3	10·5
From mid-season	5·2	5·3
Overall (grazing and conservation)	3·5	4·0

The dressings will necessarily be higher the further north one goes, but in more favourable climatic regions where clover can make a bigger contribution lower dressings will suffice. A farmer is advised to be opportunist in the use of nitrogen in order to ensure a good continuity of grazing from the early spring through to mid-autumn, as well as an adequate reserve of silage for winter feeding. Generally, silage is preferred to hay because it can be cut earlier to give a quicker and better aftermath, and there is a greater certainty of getting a reliable product, but really good hay can be just as valuable as silage for winter finishing. The difficulty is one of making it.

These are realistic targets which are comfortably achieved by the top producers recorded by the MLC. The higher stocking rate attributed to Hereford × Friesians, as compared with the straight Friesian, is principally due to their earlier maturity which means that they reach slaughter condition at an earlier age and at lower weights. They also have a lower requirement of concentrates as well as conserved grass during their last winter. Details of feed requirements in 18-month beef production from 12 weeks to slaughter are given in Table 13.2.

Supplementation with concentrates during the grazing season is usually limited to the beginning when cattle are adjusting to grazing and to the end of the season as grass quality declines and stock have to be conditioned to their winter food regime. On this account there is much to be said for feeding some silage (or hay) to cattle in the autumn before they come indoors.

TABLE 13.2. FEED INPUTS FOR 18-MONTH BEEF (MLC)

	Friesian	Hereford × Friesian
First winter (from 12 weeks)		
Concentrates (kg)	170	170
Silage* (tonnes)	0·8	0·8
or Hay (tonnes)	0·25	0·25
Grazing (6–12 months)		
Supplementary concentrates (kg)	120	80
Finishing winter		
Concentrates (kg)	630	330
Silage* (tonnes)	4	3
or Hay (tonnes)	1·2	0·9

*25% DM

Hereford-cross calves, because of their reputation, which is partly based on their flexibility as compared with Friesian or large breed crosses with the Friesian, are more expensive to buy than pure Friesians and the premium they carry at the week-old stage appears to be increasing as the incidence of Holstein breeding becomes greater. No doubt buyers argue that the Hereford will compensate in some measure for the deficiencies of the Holstein in respect of conformation. Partly for this reason Charolais, Simmental and South Devon crosses with the Friesian also enjoy a similar premium over the pure Friesian despite the fact that they are just as late maturing as the pure Friesian and require as much if not more feed to reach slaughter condition but, by way of compensation, at rather greater weights.

Farm experience and experimental work show that it is advantageous for calves to be at least 180 kg liveweight before they are turned out in the spring. This is practice means calves that are born in the period July–November. Turn-out should be a gradual process with the calves out by day and returning to the yards by night, with some continuance of previous feeding for three or four days. It is important to remember that the calves have not learned to graze properly and they will spend much of their first few days outside running around with their tails in the air. Once they are turned out completely some cereal feeding—1·4 kg daily will suffice—should be continued for 2–3 weeks when the calves should be well settled to grass and supplementary feeding can be stopped.

Hereford crosses grow about 7 per cent faster than pure Friesians but not as fast as red breed, South Devon or Charolais crosses, which have a 2–4 per cent advantage over the Herefords. The latter are the earliest maturing and it can be advantageous to have a proportion of these among a predominantly Friesian lot because a little more elbow room can be obtained for the later fattening, but bigger, Friesian bullocks in the final stages of finishing. Where the Hereford crosses finish at about 450 kg at 16–17 months, the Friesians will take another two months to reach a normal killing weight of about 510 kg usually at a time when beef prices are at a peak.

THE SYSTEM IN PRACTICE

There is the choice of starting with week-old or weaned calves of 110–25 kg or even heavier. Margins will be lower with older calves, because of the rearer's profit and additional transport costs, but they are the better choice unless first-class rearing facilities exist on the finishing farm. Moreover there is a reduction in peak capital requirements. Before the calves go out to grass they should be given the orally administered Dictol which is a reliable protection against lungworm. Though it seems an expensive treatment at the time it is important to remember that young cattle are going to be stocked very intensively and there can be a grave danger of a heavy build-up of worm infection in the late summer. A severe epidemic of husk not only means a great loss of thrift and a complete disturbance of the system, because the affected stock have to be brought inside that much earlier in the autumn, but also there is the risk of deaths from pneumonia which arises as a further complication.

It is not enough to take the above precautions against lung worm for there are also intestinal worms that can play havoc with young cattle. It is true that we now have effective drugs for the control of endoparasites but the first precaution against loss of thrift should be clean grazing, eg. maiden seeds or pasture that has been grazed by sheep or used for conservation in the previous year. Strategic drenching has its place later in the summer when cattle move to fresh grazing to deal with any infection that may have arisen despite the precautions that have been taken.

Twenty hectares of good pasture, based on ryegrass and receiving 200–50 kg of nitrogen per ha in split dressings over a season from mid-March to mid-August, should provide sufficient grazing for 70–5 calves from their turnout in mid-April to yarding in mid-October and, with controlled grazing, also provide enough silage for the following winter. It is convenient for management purposes to handle the pasture in three blocks of approximately the same area. Block A subdivided by electric fencing into four paddocks, will provide grazing on a rotational basis for the first 5–6 weeks following turn-out. Meanwhile Blocks B and C will be coming to the stage where the first conservation cut can be taken. Silage is greatly preferred to hay for a variety of reasons. It can be taken at an earlier stage of growth with an advantage in feed value, as well as an earlier and more vigorous aftermath, and there is less risk of the end-product being adversely affected by weather.

The aftermath on one of the conservation blocks should be sufficiently advanced by mid-June to provide a change of grazing and the remaining two blocks, ear-marked for second-cut silage which should be ready by the third week in July. It is prudent with this mid-June change of grazing to drench calves with a wide spectrum drench to deal with any worm infection that may have developed so that clean stock go on to clean grazing. By about the beginning of August with stock appetite increasing and grass growth decreasing as the season progresses the whole area will normally be required for grazing, though in favourable seasons it may be possible to take a third conservation cut from part of the area. The aim from the end of July onwards, however, should primarily be one of providing good grazing for the cattle so as to maximise liveweight gains from grass. It is recommended that some form of controlled grazing is continued. If each block is split in two with electric fencing to give six grazing units over the last part of the season, it should be possible to ensure that cattle have palatable grazing and waste of pasture due to treading or fouling is kept to a minimum.

SUMMER PERFORMANCE

A reasonable target is a liveweight increase of 800–900 g a day for bullocks and a little less for heifers over the grazing period, because it has been established by direct experimentation that this gives about the best compromise between individual performance and production per acre. One should not attempt to bare paddocks before each change of grazing because daily liveweight gains will suffer a probable reduction of 35–40 kg over the grazing season and this will not be recovered by compensatory growth when cattle return to yard feeding. It is important to watch the behaviour of the cattle and to act accordingly. If there is a hint of restlessness they should be moved to the next paddock because the aim must be to maintain rhythm where the cattle graze and then lie down to ruminate, rather than spend a lot of time on their feet searching for food. With mowing integrated with the cattle grazing or speaned ewes coming in as scavengers, any feed that cattle leave will not be wasted.

Records obtained by the Meat and Livestock Commission show that the most important single factor determining the level of gross margin per acre is the stocking rate per acre.

This is only true, however, if individual performance of the growing stock is not impaired to the point where output per acre starts to decline. One is dealing with much more than a simple intensity of grazing situation, for there are the interactions of levels of fertiliser use, especially nitrogen, conservation programmes and general thrift of stock. The farmers who are able to rear stock well during the grazing phase at high intensities of stocking will, other things being equal, be liberal users of nitrogen.

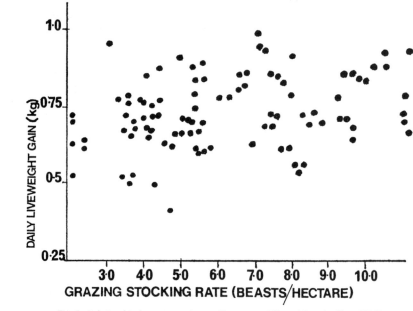

Fig. 3. Relationship between grazing stocking rate and liveweight gain (from MLC).

They will also be using conservation as a tool for preserving the quality of grazing both in nutritional terms and in respect of freedom from parasitic infections.

The normal expectation of course is that when stocking intensity is raised beyond a certain point individual performance falls, though production per hectare will still continue to rise until a break point is reached, where production per hectare and per animal both fall rapidly. Figure 3, which is a scatter diagram of daily liveweight gains at grazing in relation to stocking intensity, shows that below average individual performance is just as common under farm conditions, at low intensities of stocking of 3 beasts per hectare as it is at 7–8 per hectare.

The farmers who can achieve both high stocking intensities and good daily liveweight gains are not simply good managers of grassland. They are also first-class stockmen who are able to anticipate the need for stock to have a change of grazing before there is any manifestation of check. Under this system it is not necessary, especially in the spring and early summer, to bare a pasture to the ground in an effort to get the fullest possible utilisation before the stock are moved to their next paddock, because the conservation cut will be that much larger if pasture is not completely defoliated in the previous grazing. The main consideration in this grazing phase is to ensure that these young cattle are not on a saw-tooth plane of nutrition.

It is important not to leave cattle out too late in the autumn to grow hair and lose condition. Once the best of the autumn flush goes it is wise to introduce some rolled cereals, either 1·5–2 kg of barley or 2–2·5 kg of oats. Apart from making good deficiencies in energy intake, the cereal helps in the transition from summer to winter feeding. There is no point in feeding cereals apart from the turning out allowance during the spring and summer. In a trial at Cockle Park we obtained a response of only 0·45 kg in liveweight gain for 5·5 kg of barley and the small difference in the groups at yarding quickly disappeared as a result of compensatory growth.

SILAGE REQUIREMENTS AND WINTER FINISHING

The requirement per bullock during the winter months is about 4 tonnes of silage (of 25 per cent dry-matter content) covering a period of 150–60 days feeding. Coming in at about 350 kg liveweight, which is a good target to aim for, they will eat about 25 kg of this sort of silage plus 1·8–2·25 kg of barley daily, and by the time they are at point of finish they will be eating at least 35 kg of silage and 2·25–2·5 kg of barley. The actual level of barley fed will depend on the quality of the silage; the above allowances relate to silage cut at the ear emergence stage, when it will have a crude protein content of about 14–15 per cent and a good digestibility. More mature material cut after ear emergence will give a bigger yield per ha but intake will be impaired because of lower digestibility; more cereals will have to be fed—up to 3·5–4·5 kg daily—to achieve target gains, but at this level one has already come to the point where the silage is not good enough for feeding cattle.

Wilting of silage is recommended because this enhances dry matter intake, but it is important that the material does not become overheated because this impairs digestibility especially of protein. With well-made silage there is absolutely no need to feed a protein enriched concentrate mixture, because an intake as little as 7 kg of silage dry matter with a digestible crude protein content of 10 per cent will cover the daily needs of this class of animal with something to spare. Easy feeding is better than self-feeding to maximise silage intake.

Normally gains during the first 4–6 weeks after yarding are disappointing, and though undoubtedly some of this is accounted for by gut fill there could be other factors operating, for instance, change of diet and possibly the humid conditions that are so characteristic of November and early December. On this account it is important that buildings are well ventilated but free of draughts at stock level. Worm burdens may sometimes contribute to the poor gains, but where we have taken the precautions described previously during the grazing season we have had no response from drenching at point of yarding. External parasites can be a problem and if the cattle are constantly rubbing and are scurfy one can be very certain that they are affected by lice.

Usually after Christmas as the weather hardens, rate of gain rises appreciably and with good silage should reach 1 kg daily in January–March, to give an overall average for the second winter of about 900 g daily. Daily liveweight gains can be increased appreciably (10–20%) by growth promoting implants. The long lasting implant Compudose (Elanco) administered at turnout will give a response during grazing as well as in the final feeding period. It is important, however, to observe the withdrawal period that is specified for the several authorised implants.

COMBINATIONS OF SHEEP WITH 18-MONTH BEEF

There is much to be said for combining a sheep enterprise with 18-month beef production because it can give greater flexibility in management, particularly in respect of pasture utilisation, as well as enhance returns and give a better cash flow. This is particularly appropriate with long-term leys and permanent pasture where a change of livestock can be an important way of maintaining relative freedom from endoparasitic infections. It is stressed, however, that a sheep enterprise should not be introduced unless there is a quality of fencing that will confine the flock where it is intended that it should be grazing. Otherwise any benefits from the sheep will be more than offset by the problems they will create.

The simplest form of sheep enterprise, other than the purchase of store lambs for short-term finishing, would be the purchase of ewe lambs that are destined to be fat lamb mothers. They would arrive on the farm in the late summer just as the previous year's purchase is drafted for sale. The aim should be the purchase of cross-bred ewe lambs out of hill ewes, that are below average in size because they are twins or late lambs. Such stock do exceptionally well on clean grazing and if they are well bred they will grow into impressive sheep. Unlike a breeding flock, specialist labour requirements are small and except in the height of winter there will be little need for supplementary feeding other than some hay or silage.

A fat lamb flock, however, is capable of producing a very much higher gross margin per ha than the simpler ewe lamb enterprise and for many farmers with shepherding skills this is the more logical enterprise to combine with cattle. Working on an assumption that there are 40 ha of perennial ryegrass that is in good heart and receiving 200–40 kg of N annually one could reasonably expect such an area to provide the necessary grazing and conserved grass for 70 calves from the 6–7 months stage until slaughter at 17–19 months and 250 ewes with their lambs that are sold fat either off their mothers or within 8–10 weeks from weaning off aftermath grass.

In any production year each enterprise would have approximately the same area of grass under an arrangement where neither the calves nor the lambs graze the same pastures in successive years. The 20 ha primarily devoted to cattle would, as previously described in this chapter, cater for both conservation and grazing needs. There would be a similar arrangement with the 20 ha devoted to the flock which would be set stocked. At the height of the growing season the flock would need about 14 ha of pasture for grazing (peak stocking 18 ewes plus lambs per ha) and about 6 ha for conservation. The aftermath from this conservation area would be ear-marked for lambs remaining after weaning and these would be drenched before they went on to this clean grazing to eliminate any build up of worms during the grazing season. The dry ewe would be carried on their part of the sheep block until all the lambs had been sold and then coinciding with mating the ewes would have access to this additional grazing as well.

When the cattle move on to their finishing regime the whole 40 ha becomes available for the ewe flock so that they can be wintered at the rate of approximately 6 ewes/ha. However, if it is possible to in-winter the flock from February onwards this, among other advantages, will mean that there will be less punishment of the swards and better grazing for both sheep and cattle in the following spring. The fact that ewes have access to all the grazing from the late autumn onwards does not constitute any serious infringement of the clean grazing policy. Mature sheep pass very few worm eggs in the late autumn and any dropped in November–

January will not develop into infective larvae. It is a different matter later in the winter for, coinciding with the approach of lambing, there is the so-called 'spring rise' in the output of eggs and this also provides another good reason for taking the ewes off grazing land. This is also the time when drenching of ewes will give maximum benefit in limiting worm infections.

The above is a general outline of a plan to integrate the two kinds of grazing animal so that their thrift is maintained and at the same time there is a good pasture utilisation. In the early winter when grass growth has come to a halt, the in-lamb ewes are not under any great physiological stress, thus they can be used under a tight grazing regime to clear any neglected patches of pasture so that these are not a detriment to growth in the spring. Local conditions may necessitate some change in the numbers and balance of sheep and cattle or the levels of fertilising but this is no disadvantage provided both the cattle and the ewes with their lambs always have adequate clean pasture to satisfy their needs and there is a minimal waste of grass nutrients.

WINTER- AND SPRING-BORN CALVES

Calves born in the period December–March are not so attractive for the 18-month system as those born in the autumn. Usually they tend to be a little dearer to buy, possibly because of the demand at this time of the year for foster calves, either to replace calves that are lost in single suckling herds or for some form of multiple suckling. More important, they are not big enough to make really effective use of grass during their first summer; good gains can only be made at pasture if it is clean and free of worm infections, and some cereals are fed throughout the summer. Even at 0·9–1·8 kg daily this adds appreciably to the cost of rearing. The importance of preventing parasitic infections cannot be overstressed, and home paddocks that have previously carried young cattle are to be avoided. The safest grazing is that provided by a direct reseed in the spring, or the aftermath of a ley cut for silage or hay in its first harvest year. Advisedly this class of calf should not remain outside after the middle of September if there is any danger of lungworm.

A farmer is in something of a dilemma as to how he should handle them in the winter following their first grazing season. If cereals are cheap, which is now an unlikely prospect, it may pay to put the most forward calves on to fairly liberal cereal feeding along the lines described in Chapter 14.

The alternative is to winter them cheaply on a mainly silage diet, to grow them at the rate of 550–700 g daily so that there are good framed animals for turning out to grass. If they are Hereford crosses, and here the Hereford justifies the premium it enjoys over the pure Friesian, there is a very good chance that they will finish on grass during the course of the summer at about 450 kg and 18 months of age. The cost of finishing will be less than that with yard feeding, but the market prices for beef will also be lower than they are in the spring.

One can only expect a small proportion of pure Friesian or Charolais × Friesian calves to finish off grass at 18 months of age; the best course of action is to yard remaining cattle in October and put them on to a silage-barley diet similar to that given autumn-born calves in their second winter. They should reach slaughter condition in January–February when prices are good, at about 500–600 kg liveweight. The gross margin per head with these animals will compare favourably with that for an autumn-born winter-finished steer, but there is a slower turnover of capital and more land will be required.

ACCOMMODATION

A wide variety of accommodation is used for wintering dairy-bred stores. The principal considerations are that it should have a low cost per head, that it should provide comfort for the cattle, and be economical in terms of labour use. The covered court still has a lot to recommend it, where straw is plentiful and good use can be made of farm yard manure in the growing of high value root crops, but it is not a proposition in mainly grassland area where straw is expensive. Under these circumstances there is a lot to be said for a simple kennel system with free-draining cubicle beds to get rid of urine.

At the Grange station of the Irish Agricultural Research Institute, remarkably good results have been obtained with simply constructed topless cubicles, built in the lee of silo retaining walls. Careful comparisons have been made with both yearling and two-year-old cattle between topless and fully enclosed cubicles, and if anything the cattle have done better under the topless regime. This is a region of fairly high rainfall but not a lot of snow, but if cattle have good horizontal shelter from cold winds they show little reaction to low temperatures provided they are fully fed. There is now quite a body of Irish farmers who have erected topless cubicles, and reports on them are on the whole very favourable. Possibly they could have a value in the milder parts of Britain.

Certainly there is no justification for the most expensive wide-span buildings unless they have another essential use during other times of the year. Some of the most successful beef units occupy simple open-fronted lean-to buildings with a southern aspect. Every £100 spent on buildings carries an interest and depreciation charge of at least £16 per annum. When a producer is contemplating any additional building for his livestock it is a salutary and essential exercise to calculate the capital cost for each animal that will be accommodated and to relate this to the gross margin that this animal is likely to produce.

Chapter 14

PRODUCTION SYSTEMS

(4) Cereal Beef

DR. T. R. PRESTON, then at the Rowett Research Institute, presented a paper in December 1961 as part of a symposium on beef production that was organised by the Farmers' Club.* In it he revealed the quite remarkable results he had obtained with 20 Friesian steers which he had taken from 3 to 12 months on a totally concentrated diet of the following composition:

Bruised barley	87·5 parts
White fish meal	2·5 ,,
Soya-bean meal	7·5 ,,
Vitamin A and D	0·5 ,,
Minerals	2·0 ,,

The results he obtained were as follows:

Initial weight (kg)	93
Final weight (kg)	450
Age at slaughter (days)	362
Daily gain (kg)	1·2
Food consumption (kg)	1,433
Food conversion ratio	4·62
Killing-out (%)	54·9

It was a brief paper, presented with considerable caution, especially in respect of possible hazards such as bloat but it probably had a more profound effect on beef production than any other presentation of experimental results made over the previous 20 years. By the end of 1963 there were many farmers all over Britain who had taken up the idea; barley beef had become an established production system, which in the hands of competent operators, was also a very profitable method of producing beef from late-maturing cattle such as purebred Friesian steers.

The idea was not a completely novel one because there had been attempts to rear calves without roughage in the nineteen twenties but these failed, possibly because of Vitamin A deficiencies in the diet. More recently in 1959, Geurin and his co-workers in the United States reported good results from an all-concentrate diet,

---

* Journal of the Farmers' Club, London (Part 8, 1961).

similar to that given above, which was fed under typical feed-lot conditions over a period of 90–120 days to cattle that had been reared on grass up to 12–18 months of age. The absence of bloat was attributed to the fact that the barley was rolled, as this preserved an element of long roughage in the husks. The Rowett work was essentially an application of these American results to British conditions, using artificially-reared Friesian calves from a starting age of about three months through to slaughter at 10–14 months.

This development occurred at an important time. There was a national need to expand beef production, there was still a very high annual slaughter rate for surplus dairy-bred calves, and there was a big increase in barley production—arising from a combination of expanded acreage and higher yields due to improved varieties and higher rates of fertilisation. In the early sixties one could buy good Friesian calves for £6–8 a head, and barley could be bought for £18–20 per ton off the combine. Another development, also attributable to Preston, the early weaning system of raising calves, was simultaneously taking hold with specialist rearers; this contributed materially to the supply of suitable calves that could be purchased to go immediately on to the concentrate feeding regime.

BARLEY BEEF PIONEERS

Many of the pioneers were arable farmers and the system gave them the opportunity of cashing their barley profitably through beef. The building commitment was not a heavy one, for many farmers who had gone out of livestock had accommodation that required only minor modifications. Labour requirements were low because *ad lib* feeding from hoppers was generally adopted; the only other labour requirement apart from feeding was that required for bedding down stock except where they were kept on slats. There was much to be said for bedding the cattle on straw because they would pick it over and take in a small amount of long roughage, which was a safeguard against the bloat occasionally encountered. In the absence of straw many producers provided a small ration of hay which in practice resulted in an intake of only 0·5 to 1 kg per day, but this appeared to be a sufficient safeguard against trouble. Bloat was occasionally encountered if cattle were away from their food for some time, because they would then be inclined to eat ravenously; but with constant access to self-feed hoppers combined with the provision of some long roughage trouble from this source was minimal.

Early commercial results, with efficient producers, were remarkably similar to those obtained initially by Preston. At that time one could buy a three-months Friesian calf weighing 100 kg for about £25, which was leaving a fair margin for the rearer. The final return on the calf, including the rearing subsidy, was of the order of £80. Barley beef because of its tenderness began to command a small premium over ordinary beef because butchers found that they could obtain a high proportion of grilling or frying cuts off these carcasses. The cost of the concentrate input at that time was no more than £35 per beast, so producers were working on a feeder's margin of about £22·50 with an average investment of approximately £45 for a period of nine months. If one allows as much as £12·50 to cover labour, veterinary costs, housing, bedding and transport, then the average net return on the investment of £10 per head amounted to 30 per cent per annum. This was really big money for beef production.

It was not surprising that more and more farmers got on the band-wagon and by 1966 barley beef was accounting for approximately one-sixth of total domestic production, excluding Irish stores and culled breeding stock. It was inevitable with the growing demand for calves and the boost that barley prices received in the mid-sixties, not only as a result of its use in beef production but also because of its greater use in feeding dairy cattle as well as in an expanded pig industry, that the rate of return on capital fell to no more than 12–15 per cent, even with good standards of management. Increasingly producers have had to turn to the cheaper sorts of calves, for instance Friesians and Friesian crosses of 30–6 kg rather than 40–5 kg, and even the bigger sorts of pure Ayrshire calf, in an endeavour to maintain a reasonable level of profit.

MODIFICATIONS OF THE SYSTEM

In his original paper Preston stated that there was a need for further developments before one could confidently recommend the system as a practical proposition. There have in fact been remarkably few changes and most of these, principally emanating from further work at the Rowett, relate to levels of protein and economies that may be made in protein supplementation, which is an expensive component of any concentrate ration. Recommended feeding and management practices in the light of this further work and field experience have been summarised in Handbook No 2 of the Joint Beef Production Committee;* which gives an authoritative account of the system under British conditions.

---

* Published by the Meat and Livestock Commission, 1968.

Calves carrying a cross of the Angus or Hereford with a dairy breed are not generally recommended because they finish too quickly, and they are much more useful in a mainly grass system of production. The need is for fast-growing, late-maturing animals and this is where the Friesian and its crosses with the South Devon or the larger Continental breeds have scored.

A crude protein content of 14·5 per cent is recommended in the diet from 90–250 kg liveweight when it can be lowered to the 11 per cent level. It is important to remember that there is a considerable variation in the crude protein of barley, according to the conditions under which it is grown, especially in respect of rate of nitrogenous fertilisation. Though one normally expects a good plump sample of barley to have a crude protein of about 9 per cent it is not exceptional to have samples as high as 12 per cent; barley of this quality would supply more than sufficient crude protein for the later fattening stages, and the only additive required would be a mineral and vitamin supplement.

TYPES OF RATION

There are a number of proprietary supplements on the market, complete in respect of protein, minerals and vitamins, which can be added to the rolled barley. However, a large-scale operator would be justified in making up his own supplement because it would cost appreciably less than a proprietary one. The following supplement, which has been devised by Rowett workers, is recommended for inclusion at the rate of 3 parts of the mix to 17 parts of rolled barley (of 9–10 per cent crude protein) over the first feeding stage (90–250 kg).

Barley meal	14·6 parts
Soya-bean meal	76·2 ,,
Limestone	8·2 ,,
Salt	1·0 ,,
Vitamin A	42 million IU (per tonne of supplement)
Vitamin D	10 million IU (per tonne of supplement)
Crude protein content of dry matter—40%	
Rate of inclusion in feeding mix—15%	

Where a farmer is using a barley with a crude protein content of 11 per cent, and it is worth his while to get barley analysed for its protein content, especially if he is working on a large scale, then a supplement containing only 34 per cent of crude protein is necessary. Such a mixture could be as follows.

Barley meal	26·9 parts
Soya-bean meal	60·0 ,,
Limestone	8·7 ,,
Steamed bone flour	2·8 ,,
Salt	1·7 ,,

Vitamin A	42 million IU (per tonne of supplement)
Vitamin D	10 million IU (per tonne of supplement)

Crude protein content of dry matter—34%

In recent years with the growing realisation that the inclusion of urea in the diets of ruminants can have a protein-sparing function, it has now become a fairly general practice to substitute at least a proportion of high protein additives, such as soya-bean meal or groundnut cake, with urea which is used by rumen flora to synthesise microbial proteins which are subsequently digested by cattle. Work at the Rowett has shown that it is possible from the 90 kg stage onwards to cheapen the cost of the supplement by replacing the soya-bean meal with urea (which contains 45 per cent of elemental nitrogen or six times the amount contained in soya-bean meal) without detriment to growth rate, providing there is a compensating increase in energy and minerals, because urea only adds nitrogen to the ration.

Urea can be safely used to provide up to the equivalent of one-third of the protein of the diet in fully ruminant livestock, where there is a high level of cereal feeding as there is in barley beef production. It can be poisonous, however, if cattle get unduly large quantities of it through imperfect mixing and this is a point which must be safeguarded against if one uses the following urea-containing supplement that has been developed at the Rowett.

Barley meal	78·8 parts
Urea	6·7 ,,
Limestone	7·4 ,,
Steamed bone flour	4·4 ,,
Salt	1·7 ,,
Vitamin A	42 million IU (per tonne of supplement)
Vitamin D	10 million IU (per tonne of supplement)
Crude protein	34%
Rate of inclusion	15% of final ration

Where a farmer is feeding a barley with a crude protein of 11 per cent or more from 250 kg until slaughter, the following mixture devised at the Rowett (which has no protein additive) is recommended for inclusion at the rate of 1 part to 19 parts of rolled barley.

Barley meal	48 parts
Ground limestone	26 ,,
Steamed bone flour	15 ,,
Salt	5 ,,
Liquid paraffin	6 ,,
Vitamin A	120 million IU
Vitamin D	30 million IU

General experience has shown that rolling of barley is better

7. Ten-month Ayrshire × Charolais steers almost ready for slaughter. The value of using Charolais bulls to improve meat potential in Ayrshire is apparent.

8. A slatted floor unit used for intensive cereal beef using ad-lib feeding. Though capital outlay may be higher, slatted floors enable more animals to be kept in a given space than do conventional systems.

9. A calf nursery with capacity to house 60 animals from 4 days to 6 weeks of age. The yokes are used only at feeding times. The calves shown are Hereford crosses.

10. The utilisation of cow cubicles for winter housing of beef cattle.

than milling because milling breaks down the long fibre of the husk. Milling also results in a dusty feed which is not only less palatable, but also it is inclined to aggravate respiratory troubles. Another drawback to finely ground barley appears to be a greater likelihood of animals suffering from bloat and acidosis. Grain with less than 16 per cent of moisture tends to produce an appreciable proportion of fine dusty material when rolled, and for this reason there is much to be said for the use of moist grain that has either been stored in silos or has been treated with propionic acid.

SEX EFFECTS

Heifers of any breed or cross are less well suited than males, both entires or castrates, for cereal beef production because of their propensity to fatten before they achieve the desired slaughter weight. Steers in their turn grow more slowly but fatten more quickly than entires and consequently they are poorer food converters. Commercial records collected by the MLC give entire males an advantage of approximately 9 per cent in growth rate and 12 per cent in food conversion over castrates. Now that there has been a liberalising of official attitudes to bull beef most cereal beef producers have switched from steers to bulls. Apart from their superior growth and better food conversion bulls can be carried to higher slaughter weights without getting overfat. Friesian bulls are generally slaughtered at 10–12 months at a liveweight of 425–50 kg as compared with 400 kg for steers of a similar age. There is a marked deterioration in food conversion once prime condition is reached and it is false economy to retain animals beyond this point to increase carcass weights.

There is some evidence that an oestrogenic hormone, zeranol (sold under the trade name Ralgro and obtainable without a veterinary prescription) can be implanted in both bulls and steers to enhance liveweight gains. There is also a suggestion that it induces calmer behaviour in bulls. The implant must, however, be withdrawn 10 weeks before animals go to slaughter.

TARGETS FOR INTENSIVE CEREAL BEEF

The MLC has amassed a considerable volume of performance data from enterprise records accumulated over many years. The targets given in Table 14.1, based on these, are realistic for they are the standards achieved by a high proportion of better than average producers. Indeed they must be met by producers if there is to be a prospect of reasonable profit. Gross margins are not large, for instance the average figure achieved by producers over the 5-year period 1977–81, adjusted for inflation accordng to the General Retail Price Index, was £51 a head. This is the figure which has to cover labour, the annual cost of buildings and equipment, bank charges and a variety of other overheads.

TABLE 14.1. PERFORMANCE TARGETS FOR CEREAL BEEF (MLC)

	Bulls	Steers
Daily gain (kg)		
Week to 3 months	0·8	0·7
3 to 6 months	1·3	1·2
6 months to slaughter	1·4	1·3
Overall	1·2	1·1
Slaughter weight (kg)	430	400
Kg concentrates/kg gain		
Week to 3 months	2·7*	2·8*
3 months to 6 months	4·0	4·3
6 months to slaughter	6·1	6·6
Overall	4·8	4·5
Feed (kg)		
Milk powder	13	
Calf concentrate	155	
Protein supplement	225	
Rolled barley	1,500	

*6 to 12 weeks

The higher performance targets set for bulls underlines the importance of using entires rather than castrates for cereal beef production. Their advantage in respect of food conversion is important for food is the prime cost in cereal beef production. Typically over the whole feeding period feed accounts for 90 per cent of the variable costs and for 75 per cent of the net value of the finished animal (sale value *less* calf cost and mortality). The main thrust in any attempt to increase the profitability of a cereal beef enterprise must therefore be directed to the improvement of feed conversion. At the same time calf costs (including mortality) and the unit cost of the ration must also receive attention.

ACCOMMODATION

Accommodation need not be elaborate and it certainly must not have a high annual charge. One very successful producer of cereal beef in northeast England, which is not renowned for the clemency of its extended winters, who had an output of 1,200 cattle annually, managed with simple south-facing lean-to buildings constructed by farm labour using telegraph poles, second-hand timber and corrugated iron. It was not a prestigious layout but it was cheap and very effective. A feature of the layout was good ventilation and a minimum of draughts at stock level. This is important because cattle bedded on straw are able to lie in comfort. It was also a very accessible layout for feeding and removal of manure and, not least, when the farmer decided to go out of cereal beef there was a relatively small amount of capital tied up in the redundant buildings.

Temperature in itself, with the normal range that is encountered in Britain, does not greatly affect the performance of housed stock, provided there is ample fresh air to reduce the humidity level. High humidity leading to condensation affects the fabric of a building, it results in damp bedding and it can accelerate the spread of respiratory diseases. These ill effects are particularly marked in the late autumn–early winter period of mists and fogs. It is a common experience with feeders, when cattle are brought into yards in the autumn, that they make only moderate gains until Christmas. This has been variously accounted for by such factors as change of diet, parasitic infections and change of environment, but it is also experienced with cattle on intensive cereal feeding where there is no change in the management regime.

Figure 4, which is based on records collected by the Beef

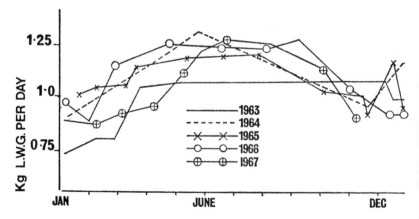

Fig. 4. Average daily gain of cattle on intensive cereal feeding systems 1963–67 according to month of the year (Handbook No. 2, MLC).

Recording Association over five years, shows how substantial and regular this decline is under farm conditions. Though one cannot be categoric that high humidity is the only operative factor, because a light factor could also be involved, undoubtedly it is very important because of the substantial improvement in stock performance that occurs when ventilation deficiencies are corrected. These need not be elaborate measures, for example in an open-fronted lean-to house, preferably facing the south, an opening along the back wall at eaves level combined with Yorkshire boarding of the gable ends, above stock level, will be adequate. Where there is a pitched roof there should be ridge ventilation. If in the autumn there is any appreciable incidence of coughing in cattle that have been continuously housed, then one can be very certain that ventilation is at fault.

It is important to try and keep groups of cattle together from the early rearing stage through to slaughter, because they settle to a social order which is disturbed when lots are mixed. Slatted floored accommodation is rather more economical in space requirements per animal than bedded areas, eg. 2·4 m² for a beast of 225 kg or more on slats as opposed to 3·3 m² of bedded area. As little as 1·2 m² on slats or 1·5 m² on straw will suffice for an animal of 100 kg, so that in order to economise in space requirements it is advisable to have pens of differing sizes so that groups can be moved to larger pens as the need arises. Animals within groups should be matched for size, and it is preferable to have no more than twenty beasts per pen.

Cattle appear to do just as well on well constructed concrete slat floors as they do on straw. Slats 130 mm wide with 38 mm spacings are recommended and they should be laid with the long axis parallel to troughs and hoppers. Where hay is being fed from racks there is a danger of pulled out material impeding the movement of dung through the spaces and so it is preferable to use a Scandinavian hay box with a wire mesh cover to minimise this nuisance.

These remarks concerning accommodation apply to other forms of indoor finishing as well as barley beef. Generally, however, it is necessary to have different shaped pens to ensure that there is sufficient trough space for all the animals in the pen, a consideration that does not apply with cattle that are on *ad lib* feeding from hoppers. While 380–460 mm run of trough is ample for animals up to 225 kg liveweight at least 760 mm is required for beasts of about 500 kg.

One essential piece of equipment for all systems of feeding is a weigh-bridge, because without the guidance on performance that weighings provide a feeder is working largely in the dark, except for eye observations which are sometimes misleading. It is not necessary to weigh every animal once a month, but sample pens representative of the various stages of fattening will give sufficient information about the progress of stock up till the last month before slaughter, when all animals should be weighed at fortnightly intervals.

GRASS–CEREAL BEEF

A variant of the straight cereal beef system described above is the rearing of calves on grass for periods of up to six months, following a normal early weaning regime and then finishing them on a mainly cereal diet at 12–14 months of age. It is more appropriate to December–March born calves than to earlier born calves which fit in better to the 18-month system.

These younger calves are rather more difficult to manage on grass in their first summer as compared with autumn-born calves, and one is doing well to get a gain as high as 700 g per day from grazing. It is vitally important that the calves are grazed on clean pastures, eg. maiden seeds or aftermaths, with fairly frequent changes. They should not be left out in the late autumn to grow hair rather than put on flesh, and so it is advisable in most districts to house them by the first week of October.

The transition from summer feeding to the winter regime must be gradual. Rolled barley should be fed at the rate of 1·4–1·8 kg daily during their last fortnight on grass, and when the cattle are brought inside concentrate feeding can be gradually stepped up to appetite, using a conventional cereal beef diet of the kind that has previously been described, with hay on offer all the time. Once the cattle are over the transition period they should be capable of putting on at least 1·1 kg daily; if they are Friesian or Friesian crosses they should reach slaughter condition at 410–30 kg by liveweight at a time of the year when beef prices are high. This last point is the principal justification for putting such cattle on to a cereal rather than a roughage plus cereal ration of the sort used for finishing 18-month beef, because there is a reasonable certainty that they will reach the required degree of finish in time to catch the market while prices are at their highest.

However one would recommend it primarily as an opportunist operation to be undertaken only if barley is relatively cheap in the autumn. Such are the prices of good stores of 12–15 months of age in the spring, as a result of over-optimistic buying by grass feeders, that it will generally be more profitable to winter this class of cattle as cheaply as possible on a silage and/or hay diet, with very limited concentrate supplementation, to grow frame rather than to feed them on a full cereal finishing ration. Especially is this likely to be the case with Hereford or Charolais × Friesian stores that command a premium over straight Friesians for grass feeding.

Chapter 15

PRODUCTION SYSTEMS

(5) Fattening on Grass

BETWEEN THE wars grass finishing of cattle was sometimes described, in a rather deprecating way, as dog and stick farming but nevertheless it was an operation that required considerable skill and judgement. There were permanent pastures in several parts of Britain that were renowned for their feeding qualities and these were mainly located on inherently fertile, strong land with ryegrass and white clover as the dominant species. Management of these pastures including levels of stocking was according to a pattern established over many years. There was little in the way of fertiliser applications, possibly an occasional phosphate dressing with basic slag as the principal fertiliser but never nitrogen because apart from cost this would have been to the detriment of clover which was prized on two accounts: its contribution to the nutritive value of the pasture and its value in fixing atmospheric nitrogen to sustain grass growth. Traditional grass fattening had to be a low input system of land use and in this sense white clover provided something for nothing.

MIDLAND FATTENING PASTURES

Nowhere was the art of exploiting fatttening pastures better demonstrated than in the Leicester–Warwick–Northampton triangle known as 'The Shires' where grass finishing of cattle has considerable antiquity. There was strong resistance in this area to the plough-up policy of World War II on the grounds that it would take a very long time to recover the intrinsic fattening qualities of these age-old permanent pastures when the land was ultimately returned to grass. Feelings on this question were so high that the Royal Agricultural Society of England sponsored a series of trials to compare leys and permanent grass. These were not confined to the Midland region because it was considered important to extend the comparison to cover different qualities of permanent pasture on a variety of soil types.

The final report* established that permanent pastures compared favourably with leys only when they were of the highest quality. This was no more than one could expect for at that time the majority of permanent pastures in Britain had very little white clover and a dominance of the less productive grass species such as Agrostis and Yorkshire fog and in this respect they differed markedly from the better Midland pastures. The leys that replaced these low-grade permanent pastures were mainly based on improved strains of rye grass and white clover and they were much better treated, particularly in respect of fertilising and liming, than the poorly-managed permanent pastures at the lower end of the production scale. But it was a very different story at the top end, so much so that it is doubtful whether Midland graziers and other farmers with comparable pastures througout Britain, derived much comfort from the report as they contemplated land in tillage that had once been their pride and joy as top quality feeding pastures. There was substance to their viewpoint for many years of good management of permanent pastures on inherently fertile land had undoubtedly created dense leafy swards with an extremely good mid-summer performance. Most farmers with experience of both leys and permanent grass used for fattening agree that the latter gives a more sustained production of leaf than leys, particularly with set stocking.

Traditional grass feeders always favoured older bullocks, particularly on their better pastures on the grounds that younger animals tended to scour and lose bloom whereas 3-year-old bullocks would fatten readily. They also favoured set stocking rather than any form of controlled grazing because of a disturbance factor. Cattle on controlled grazing tend to become restless as pastures are eaten-out prior to their next move and it is recognised that stock must maintain a steady rhythm of grazing, cudding, and resting if they are to give of their best. This may lead to an apparent under-utilisation of pasture, as compared with a dairy farm where grazing is controlled and grass surplus to immediate grazing needs is conserved to give clean aftermaths with new growth. Graziers with long experience of the capacity of their pastures were able to judge to a nicety the best intensities of stocking over the grazing season to maximise individual animal performance as well as sustain a high level of production per unit area. This was part of the art of managing these fattening pastures to best advantage.

---

* Davis, William and Williams, T. E. (1954), *Journal of the Royal Agricultural Society of England*, vol 115.

Most good feeding pastures were capable of finishing two lots of cattle over the course of the season to give a net output of the order of 300 kg/ha of liveweight gain which incidentally is half of what one can reasonably expect from steers intended for 18-month beef during their spell at grass.

It was inevitable with this style of stock management that at the end of the grazing season pastures would be a mosaic of closely grazed and neglected patches. Graziers preferred not to bare them with more cattle grazing because of the dangers of poaching on strong land often with a high water table. This was where a flying flock, sometimes store lambs and sometimes two-tooths were brought into the operation as animated mowing machines which also made their small addition to farm income. Their primary function, however, was one of removing rough growth so that there was a clean fresh pasture to greet the next spring's intake of cattle.

CURRENT PRACTICE IN GRASS FATTENING

This older style of grass finishing is now practically little more than a memory in Great Britain because of several factors. For instance the supply of 3-year-old steers has practically dried up with the virtual disappearance of imported Irish stores which were once greatly favoured for grass fattening. Indeed 2-year-old cattle, from any source, are also becoming thin on the ground for the simple reason that there is now so much capital tied up in a beast that it makes better sense to realise on this investment as quickly as possible. There is another factor also contributing to such a policy, namely the greater efficiency of food utilisation that comes with a reduction of the feeding period. This important lesson was one of the benefits of barley beef production where there is no store period. This is only justified where some cheap form of nutrient can be utilised to grow frame in readiness for finishing on a seasonally available and comparatively economical diet such as grass.

Grass finishing in Britain is now principally concerned with two classes of cattle, suckled calves that are too young or have insufficient size to justify any effort to sell them fat out of yards at the 15–18 month stage and dairy-bred stores which are mainly winter- and spring-born and therefore, unlike autumn-born calves, are not well suited for the production of 18-month beef. Regardless of source of stores, however, there are many lessons deriving from the traditional feeders' approach that continue to be important, and possibly the most important is the necessity for sensible buying. The old hands were masters at this game and they needed to be, for then the feeder's margin was only a few pounds and in those hazardous days, when there were no guaranteed prices, unwise buying could soon land a grazier in the bankruptcy court.

Possibly as a result of many years of guaranteed markets with regular annual rises in beef prices the store cattle markets in the spring sometimes gives the impression that they are patronised by a race of super-optimists. In one respect it is reasonable to pay rather more per kg of liveweight when buying store cattle, particularly if they are on the lean side because their potential for gain usually includes something extra in the shape of compensatory growth. Conversely, however, cattle winter in covered accommodation and carrying a lot of condition lose flesh when they go to grass in the spring, especially if there is typical April weather. For this reason old time graziers were usually prepared to pay a little more for cattle that had obviously been outwintered since they would put on condition as soon as grass came away in the spring.

Table 15.1 based on MLC records gives some indication of the relationship between the liveweight cost of stores and the selling price of grass finished cattle over the 5 years 1977–81 together with grass margins per head.

TABLE 15.1. GRASS-FED CATTLE, PRICES AND GROSS MARGINS, 1977–81

| | Cattle prices (p/kg liveweight) | | Gross margins/head (£) | |
	Store	Sale	Actual	Adjusted*
1977	60	56	38	82
1978	72	67	46	70
1979	82	75	52	68
1980	92	79	37	41
1981	100	92	42	42
			5-year average	59

*Adjusted for inflation according to General Retail Price Index

Over the period store prices averaged 7p per kilo more than fat stock prices with the difference being as much as 13p in 1980 which was also the poorest year for gross margins which had declined progressively from 1977.

Buying of store cattle is not something that can be lightly undertaken by beginners and unless one is prepared to spend a lot of time at markets getting a feel of values it will invariably pay, as it does with the purchase of rearing calves, to commission a reliable dealer to make the required purchases. These men not only have a knowledge of prices but also judgement in respect of potential. The price of achieving their expertise may be the cost of many expensive mistakes and there is the safeguard that an established dealer knows very well that if he is to retain the goodwill of his clients he cannot afford to let them down.

A dealer's services are especially valuable in the spring when most farmers can ill afford to be away from their farms chasing elusive bargains with the pressures they have to face at this time of the year. Autumn, the season of suckled calf sales, is usually a less demanding time and farmers are more likely to buy stock at sensible prices. Provided a grass fattener has the necessary fodder and in-wintering accommodation there is much to be said for buying from the lower end of suckled calf offerings, which owe their lack of size to age rather than their breeding or ill thrift for this class of stock is not in demand for winter finishing. The aim with such stock is to grow frame rather than flesh during the winter and a daily gain of 0·5–0·6 kg is a reasonable target that is attainable with average quality of silage fed to appetite. There is some advantage in wintering them in semi-open yards so that they are used to having weather on their backs because they will then be less likely to suffer a check when they go out to grass.

Autumn is also a time when considerable numbers of dairy calves come on to the market as rising yearlings and these can sometimes be a better proposition than suckled calves for wintering and subsequent grass finishing. MLC records show that they usually cost less per kilo than sucklers but they also sell for less per kg liveweight. Late-maturing animals such as Friesians or its crosses with any of larger breeds often fail to reach slaughter condition on pasture at the end of their second summer, unless they are given a cereal supplement as the nutritive value of grass declines with the shortening of the days. The alternative is yard finishing on silage and cereals to take advantage of the rise in the prices that normally occurs with the onset of winter. Hereford × Friesians on the other hand will invariably finish on pasture if it is of fattening quality but one has to pay appreciably more for them as stores on this account.

MANAGEMENT AT GRASS

The primary aim in grass feeding, particularly early in the season, is less one of maximising output per hectare than maximising individual performance and, as pointed out earlier in this chapter, set stocking rather than any form of controlled grazing is favoured on this account. About 50 years ago Professor Johnson Wallace, a pioneer in behavioural studies, investigated the behaviour of cattle grazing pastures at contrasting stages of growth. The optimum appeared to be the 100–20 mm stage (4–5 inches) which incidentally is a highly nutritive stage. Cattle spent no more than 8 hours actually grazing with the remainder of the day resting and cudding. With very short pasture they can spend up to 12 hours on their feet either grazing or looking for food. At the other end of the scale on long pasture with its lower nutritive value cattle again spend a lot of time grazing or fossicking at the expense of resting and cudding, the normal behavioural pattern of contented stock.

Early in the season it is more than a matter of getting good rates of gain, either from the individual animals or the pastures they graze. From about the end of July there is normally a substantial fall in beef prices responding not only to increased supplies but also to competition from an influx of lambs on the market. It is in a grazier's interest to catch the market for fat cattle before it goes into this seasonal decline. This is where early-maturing cattle have an advantage over later-maturing types but unquestionably quality of grazing is also of prime importance. A liveweight gain of $0.8–0.9$ kg per day over a period of 140 to 150 days would be regarded as a satisfactory performance.

The successful grass fattener walks a tightrope in a sense that he tries to maintain the optimal balance between stock performance on the one hand and pasture utilisation on the other, and he has to do this balancing act with seasonal changes in grass growth. Throughout April a bunch of 25 bullocks may require as much as 20 ha of pasture to satisfy its needs but by mid-May 10 ha will be sufficient. A grazier can handle this situation either by drafting in more cattle to increase grazing pressure or by concentrating the existing cattle on a reduced area with the balance being laid up for conservation. If the latter course of action is adopted then it is better that the crop is taken early for silage rather than later for hay for this not only means that the aftermath will be earlier and more leafy than a hay aftermath but also that there will be reserve grazing immediately available as a safeguard against summer drought. This aftermath is especially valuable where the cattle are mainly late-maturing sorts for they maintain grazing pressure until the end of the season.

At the expense of the pleasing sight of a uniform lot of cattle on one's farm there is something to be said for having both suckler-bred and dairy-bred stores because this would give a better equation between stock appetite and availability of grass over the year. It is good sense to give the early-maturing cattle preferential treatment so that they can be sold before grass growth starts to decline. Heifers are also useful in this respect because of their early maturity but they must be grazed separately from bullocks because of sexual activity.

NITROGENOUS FERTILISING

No mention has yet been made of fertiliser practice. Most graziers will ensure that there is no shortage of phosphate, potash or lime but they tend to be sparing users of nitrogenous fertilisers in the belief that if there is a significant loss of clover there will be a deterioration in the feeding quality of the pasture. This is important where a farmer aims to sell the spring intake of stores not later than the end of July. Despite this a March dressing of 40–50 kg N/ha is recommended to boost growth for the hungry month of April and it will have little effect on the clover content of the pasture later in the season provided it is kept under control in May when grass is normally the dominant component of the sward.

Nitrogen is of greater significance where there are later maturing stores of the kind that either need a barley supplement in the autumn if they are to finish off grass or are destined for yard finishing in the early winter. Here the situation, certainly in the early part of the season, is more akin to that with 18-month beef where one is primarily concerned with growing frame and maximising output per ha. Here there can be good grounds for an opportunist approach to the use of nitrogen. If grazing pressures are such in relation to availability of grass that a boost is required then judicious applications are valuable but if there is abundant grazing then it is better that the fertiliser remains in store.

Nitrogen of course is particularly valuable where pasture is taken out of grazing and laid up for silage with the ultimate intention of grazing the aftermath. Here a combination of nitrogen and potash or a complete fertiliser such as 20–10–10 is recommended rather than straight nitrogen. Apart from the replacement of P and K removed in the forage a deficiency of these elements can limit clover growth.

SHEEP IN ASSOCIATION WITH FEEDING CATTLE

Some farmers favour a light stocking of ewes and lambs along with fattening bullocks partly on the grounds that lambs do extremely well on such grazing and partly because this may prevent pasture getting too rank. This may be a satisfactory arrangement on a farm where there are relatively few sheep but it is a different matter with intensive fat lamb production. Here it is better to adopt 'a clean field' programme where ewes with lambs alternate annually with the cattle in the use of a given pasture for reasons described in Chapter 13.

Dry ewes, however, can be used to good effect in the winter

months along the lines adopted by the old-time Midland graziers to trim pastures in readiness for the next season. Neglected patches of rank pasture, usually due to fouling by excreta, are inevitable under a system of relatively lax grazing that benefits cattle performance but will cause deterioration of the sward unless they are eaten out. This is a job in-lamb ewes can do in December and January without detriment to their well-being and without any meaningful infection arising from the deposition of worm eggs.

ECONOMIC CONSIDERATIONS

Table 15.1 is a reminder that feeding cattle on grass is seldom a very profitable venture. The 5-year average gross margin of £59 per head must be related to an investment of the order of £350 in a bullock for a period of at least 5 months. A farmer working on an overdraft could have bank charges of the order of £25 a head and the margin per beast to cover rent, labour and other overheads could come down to £30–5. It seems that there can be precious little left after meeting these charges that can be regarded as profit. It is pertinent to ask therefore whether grass fattening is really a viable proposition when other forms of land use are much more profitable.

There can be no categoric answer either one way or the other but it must be remembered that the performance of the most successful operations is very different from that of the average. Their judgement in buying stores and marketing their fat stock combined with their skill in managing stock at grass can return a reasonable profit with a much lower labour input than that required for any other class of stock. For somebody who has been dairying most of his working life acceptance of a golden handshake and a switch to grass fattening must seem like a gentleman's life.

It must be remembered also that few farmers even in the traditional fattening districts now have grass feeding as their sole enterprise. Usually it is ancillary to some more important enterprise, because the plough that transformed Midland fattening pastures into tillage has remained as an integrate in a widespread system of alternate husbandry where cash crops, particularly cereals, are the principal source of income. Grass finishing is only one of several means that farmers have for exploiting their grassland. Dairying, 18-month beef and intensive fat lamb production are all potentially more profitable than grass finishing though the combination of over-wintering stores for subsequent finishing on grass can give gross margins that compare favourably with those from alternative enterprises, apart from dairying, with many less worries as well as a much lower labour input. This, as

MLC farm records reveal, is the better approach to grass fattening not only because there is a profit to be taken from overwintering and another from feeding but also the hazards of spring purchases are avoided. Then there is the aesthetic aspect for there are few more pleasing sights in farming than quality cattle on a good pasture that are coming up to prime condition. Regrettably, however, it is the plainer sort of cattle that usually make the most money and it seems that a farmer is well advised to pocket his pride when buying store cattle.

Chapter 16

MEAT AND CARCASS QUALITY

QUALITY, LIKE BEAUTY, is a very subjective attribute which varies from country to country and region to region. Various definitions have been put forward over the years, but all have suffered from the lack of any objective approach and have generally concluded that quality meat was that for which the public was prepared to pay the highest price.

DEFINITION

Before World War II the Englishman's ideal of quality was, without doubt, exemplified by the kind of meat one could expect from a well-finished Angus steer: a bright cherry red lean, hard white fat and a high degree of marbling. Of course, in those days fatty meat was not held to be undesirable, but since the war there has been a gradual swing away from extreme fatness towards an almost excessive leanness. British housewives are still personally advised as to quality by their local butcher, but this is becoming less so with the increase of supermarket meat counters, while in Continental Europe the housewive is influenced by quite different motivations from those of her British counterpart.

Bearing in mind the dangers of attempting to generalise upon a trait as subjective as meat quality one can assume that there are three basic consumer outlets. These are the processing industry, the general trade (housewife) and the luxury trade. Clearly all three have quite different requirements and hence ideas of quality, in terms of what each outlet is prepared to pay most for, are also distinct.

The processing industry places leanness above all else. It is probable that no other considerations of eating quality are considered other than this absence of fat. The housewife, on the other hand, while still requiring a minimum of fat, will be influenced by other aspects such as good coloration and, eventually, with eatability. The luxury trade, which covers hotels, high-class restaurants and the like places greatest emphasis upon eating quality and still retains the pre-war ideals of what constitutes that quality by

181

seeking well-marbled meat with a substantial fat cover in the belief that this will lead to better flavour and juiciness.

Since the luxury trade is inevitably a minor one, the beef producer will primarily be seeking a carcass with maximum lean content and, since no payments are or can be made on eatability, will pay little attention to other aspects.

LEAN MEAT

The amount of lean meat that an animal will produce will depend upon the size of its carcass and the composition of that carcass. Body size and carcass weight are related in a positive sense in that the heavier the animal the higher will tend to be its killing-out percentage. This later trait is not only related to body weight but also to carcass fatness. In general, at any given liveweight the fatter an animal the higher will be the killing-out percentage. This rule will generally hold good within any breed though it will not necessarily hold across breeds. For instance Limousin crosses have a higher killing-out percentage than the cross-bred progeny of any other beef breed used in Britain despite the fact that they produce very lean carcasses with an exceptionally high proportion of saleable meat. Dietary systems will also have a role to play in that bulky diets will tend to produce lower killing-outs at the same liveweight than will concentrate diets, mainly due to differences in gut fill.

Thus the higher the liveweight and the more intensive the diet, the higher will be the carcass weight obtained. This does not, of course, mean that one should kill at the highest possible weights in order to maximise lean meat since this character will not react in the same way. With cereal beef systems, liveweights will tend to be of the order of 400 kg since this will produce acceptable carcasses with maximum lean yield, provided that late maturing breeds are used. The traditional beef breeds are much earlier maturing than Friesians or European beef breeds and under barley beef systems would tend to be too fat by 400 kg. In contrast 18-month beef systems based on pasture and pasture products will produce less fat in the carcass, hence liveweight can be correspondingly higher than with cereal beef. Here again breed effects will be important, since early maturing breeds will have to be slaughtered at lighter weights than late-maturing breeds if lean percentage is to be maximised, albeit at higher weights than would be possible with cereal beef.

Sex will also play a part in lean meat yield, since at any given carcass weight, bulls will be leaner than steers which in turn will be leaner than heifers. Fatness, since it is the converse of leanness will

11. Zero grazing of Friesian steers. This unit comprised 113 head on grass from 13·5 ha. Over a 134-day period starting April 10 the steers averaged 770 g per day liveweight gain from an initial weight of 178 kg.

12. A Sussex bull, *Fletching Cassius*. Growth-recorded at a BOCM trial, this bull weighed 526 kg at 365 days of age. He was placed first of five on conformation. Although performance testing techniques still need refinement they represent the main hope for genetic progress in beef cattle.

13. A polled Hereford bull, *Knightwick 3 Goldsmith*. This animal performance tested at Holme Lacy and weighed 550 kg at 400 days of age. This was the highest weight in the group of 19 and was 71 kg above average.

14. Groups of Hereford × Friesian steers being fattened at Warren Farm as part of a progeny test of their sires. Genotype-environment interactions were studied with progeny being fattened on grass or as cereal beef.

show the reverse trend, while bone will tend to follow similar patterns to lean though less pronounced.

The beef producer must determine slaughter point on the basis of the feeding system, sex and breed as well as market prices, knowing that total lean meat will decline with increasing size (in percentage terms), fatness will increase, bone will decline and that first quality meat (as a percentage of total meat) will remain fairly stationary, ie. it will decline as a percentage of carcass weight. Since percentage bone declines more rapidly than lean, the meat:bone ratio will increase with increasing liveweight and the lean:fat ratio decline.

ASSESSMENT OF LIVE AND DEAD CATTLE

Appraisal of a live animal for its carcass qualities is not a very precise operation. Apart from the use of the Scanogram, an ultrasonic meter that can provide reasonably accurate measurements of subcutaneous fat depth over the *longissimus dorsi* (eye muscle) there are no linear measurements that have the required reliability to give these any meaningful place in the evaluation of live animals. Apart from the use of scales for determining weights, live animal appraisal is essentially subjective and even experienced judges cannot assess the lean content of resulting carcasses with any degree of accuracy. Indeed their tendency to give preference to blocky, smooth-fleshed cattle may result in their choice resting on animals that carry too much fat for present trade requirements.

The most that the average producer can hope to do is to judge whether animals are ready for slaughter. This is important for carcasses can be down-graded if they have insufficient finish or if they carry too much fat, while over-fattening constitutes an unnecessary waste of food. A farm model of the Scanogram could be very useful but it is still a tool of research rather than commercial production.

Killing-out percentages are an important consideration with buyers of fat stock and experienced butchers become very adept in making judgements provided they have some knowledge of the diets of the cattle they are buying. An ability to determine carcass yields of animals, bought on the hoof, with a high degree of accuracy is immensely important for large scale buyers operating on behalf of such organisations as frigorificos and packing houses. In fact if they lack this necessary experience they will be unable to hold down their jobs for very long. An error of 1 per cent in 50,000 beasts represents a lot of meat and a substantial loss if the error favours the seller. For this reason, along with others, sale of animals on the hook has much to commend it. Provided there is a satisfactory system of payment on grade, it brings home to producers any deficiencies in the stock they have offered for sale.

Grading of carcasses is an important consideration not only for the producers selling animals on the hook as opposed to the hoof where the buyer has to stand by his judgement, but also to the wholesale and retail meat trade which is progressively being concentrated in fewer hands. Except in rural districts the family butcher who buys fat stock in a local market and slaughters them in his own or a small local authority abattoir is becoming a rarity as more of the nation's retail trade in meat is taken over by chain butchery concerns and supermarkets. These large concerns do not buy penny packets, for much of their competitive strength derives from the economies of large scale buying. Their purchasing agents like to order according to meaningful specifications as they are able to with New Zealand lamb or Danish bacon which are both offered for sale with well-defined grades in which buyers have complete confidence.

The development of an acceptable system of beef carcass classification was, with considerable justification, a priority task for the Meat and Livestock Commission and, on a basis of knowledge accumulated by the dissection of carcasses and part carcasses MLC developed a system of grading that was sufficiently reliable to be accepted by an encouraging proportion of the meat trade. Meanwhile within the EEC, there was a growing realisation of the importance of having a universally acceptable system of carcass classification and since November 1981 Britain has adopted the EEC carcass classification scheme that owes a lot to the pioneer work done by the MLC. This is designed to facilitate dead weight price reporting and intervention purchasing on a comparable basis throughout the Community.

This scheme, like the one it replaces, classifies carcasses according to their conformation and fatness. Its basic features are shown in Fig. 5. Conformation, going from very good to very poor, is classified by the letters E, U+, U, R, O, O−, P and increasing fatness is classified numerically. The system appears to be acceptable to a somewhat conservative trade for in 1983, 29 per cent of the cattle slaughtered in Britain were classified under the EEC Scheme. In view of the continued emphasis on auction selling of fat stock this is encouraging. An important feature of the classification is that it does not attempt to establish relative merit in respect of the different sorts of carcasses and there are no evocative terms like prime or choice. Its purpose, sensibly, is to classify carcasses to enable purchasers to make the choice that best suits their particular trade requirements.

Conformation	Fatness (from very lean to very fat)						
	1	2	3	4L	4H	5L	5H
Excellent E			E3				
U+							
U							
R				R4L			
O							
O−							
Very poor P		P2					

Fig. 5. EEC classification of carcasses according to fatness and conformation.

Conformation is quoted before fatness in classification, for instance an average sort of carcass could be described as a R4L and a lean carcass with a very poor conformation could be classed as P2. Carcasses in the E conformation category are very rarely encountered in Britain. Carcasses with the double muscling characteristic of some strains in a number of Continental breeds (notably the Piedmont cattle in Italy and the Blonde d'Aquitaine, one of the breeds imported into Britain) are likely candidates for the E classification which accounted for less than 0·3 per cent of the 1982 classification.

MLC data relating fat class to sex confirm that young bulls are much leaner than steers which in turn have less fat than heifers. During 1982, 68 per cent of young bull carcasses had ratings of 3 or less. The proportions for steers and heifers were 23 and 17 per cent respectively. There were no meaningful differences in conformation classification between the sexes with over 80 per cent of each category coming into the middle of the road U, R and O ratings. The cross cattle by imported Continental breeds have an appreciably better conformation rating than those by British breeds when comparisons are made under uniform conditions.

EYE MUSCLE AREA

The *longissimus dorsi* is the most important muscle in any meat carcass because it constitutes the eye of lamb and pork chops and it is the main lean constituent of sirloin and rib roasts. Measurements of a cross section area of this muscle in the loin region have been used in attempts to establish equations that predict the lean content of a whole carcass but these have been less than moderately successful. Nevertheless area of eye muscle is important in its own right, particularly with the increasing demand for grilling and frying cuts, such as sirloin steaks, at the expense of roasts and other large joints.

Eye muscle area has a reasonably high heritability (Preston & Willis* quote eight estimates in the range 0·4 to 0·76) and there are appreciable differences not only between breeds but also within breeds. Eye muscle area is closely related to the overall size of an animal, for instance 30 cm² in a 100 kg Hereford as opposed to 90 cm² in a 500 kg beast. Unlike the position with pigs, eye muscle areas in cattle cannot be measured with accuracy in the living animal using ultrasonic meters. On this account greater genetic progress is likely to be made by progeny rather than performance testing, though this will inevitably mean an extension of the generation interval. It is necessary, of course, to have a standardised rearing routine with test animals being slaughtered at the same weight if the progeny test is to have any validity.

ATTRACTIVENESS

Components of attractiveness are colour of lean and fat, firmness, texture and blood drip. Lean colour is generally considered to be best when a bright cherry red and although a wide range of acceptability exists, dark cutting beef is undesirable. This is related to meat pH in that normal meat is about 5·6 pH and at 6·5 or more it becomes dark. Although some breeds such as the Brahman tend towards darker cutting beef these are not present in Britain, although there are suggestions that Friesians may give darker meat than traditional beef breeds. In this country the major cause of dark cutting is stress of the animal prior to slaughter because this brings about muscle glycogen loss which in turn raises pH. The problem is important in pigs which have lower muscle glycogen reserves, but in beef cattle is only brought about by excessive stress and almost continual goading. In general colour tends to become darker with age, and this may be of some importance with bull beef which might be darker than that of steers, especially under grass-feeding conditions. Under normal conditions dark cutting beef is unlikely to be a problem. Unless badly stored or a product of old cull animals, most beef is likely to fall within the required colour range.

The same is true of fat colour, where only in the case of those crosses with the Jersey or Guernsey could fat colour be expected

---

* Preston, T. R. and Willis, M. B. (1970), *Intensive Beef Production*, Pergamon Press, p. 119.

to show a marked yellow coloration and thus prove less acceptable. This coloration is due to vitamin A storage and hence prejudice is aesthetic since the fat is actually of higher nutritive value.

FIRMNESS AND TEXTURE

Firmness of flesh is more a problem of pigs in which watery muscle is receiving considerable publicity. It has been associated with pre- and post-mortem treatment of cattle and may be connected with water-holding capacity. Firmness does not seem to be associated with fatness, and well-marbled carcasses are unlikely to suffer from watery muscle.

Texture is a trait that is difficult to assess before purchase while after eating complaint is futile. It does seem that texture is related to connective tissue and hence coarse-textured meat will be tougher to eat. There is some evidence that bull meat is coarser textured than that of steers, and that animals of Charolais origin may produce coarser meat than those of the traditional beef breeds.

Blood drip is unsightly on the butcher's slab and hence mitigates against ready sale. It is alleged to increase with carcass fatness and is greater in frozen than fresh or chilled beef. Some suggestion that increased pH would reduce drip is of little practical value since increases in pH are antagonistic to colour and tenderness as well as flavour.

TENDERNESS

The most important single trait affecting eatability is tenderness. It is also the easiest to measure, either by use of shear machines which measure the force needed to sever a core of cooked meat or by panel studies. In this latter connection it should be remembered that panels are trained observers, and as a result they can frequently discern differences that consumer acceptance studies cannot. Taste panels have been known to identify differences between sire groups whereas some consumers would have difficulty distinguishing between species let alone sires.

Tenderness is thought to be related to the amount of soluble collagen in the muscle. Other aspects can be important, hence cutting muscles from the carcass prior to rigor mortis will increase toughness, while ageing of the meat for about thirteen days or intravenous injections of papain immediately pre-slaughter will improve tenderness.

It is established that tenderness declines with age and that this is particularly true with bulls as opposed to steers or heifers. Bulls will generally be less tender than steers, but under 400 days this is less obvious and may be impossible to identify in cereal-based diets. Hormones have no apparent effect on tenderness.

It has been long believed that marbling is an important indicator of tenderness, but in fact only about 5 to 11 per cent of the variation in tenderness is attributable to marbling effects. In extreme cases marked relationships do occur; comparisons between well-marbled carcasses and those devoid of marbling will demonstrate greater tenderness in the former. Over more normal ranges, however, the relationship is tenuous. Cooking does improve tenderness by hydrolysing collagen of the muscle and there is some suggestion that more hydrolysing occurs in fatter areas of the meat. However there is no conclusive evidence on this issue, particularly since some interaction between cooking method and marbling exists. For example, as regards tenderness, marbling has been found to be more important in braised rounds than in grilled loin steaks.

Higher planes of nutrition will increase fatness though not necessarily marbling, but they will improve tenderness and it has been shown that grass fed animals, while leaner than those fattened intensively, gave tougher steaks with greater cooking losses.

There are distinct breed effects on tenderness but these have often been overemphasised. It is frequently claimed that Angus and Herefords have better eating quality, but as regards tenderness US work has shown similar tenderness values in meat of Friesians and Herefords on similar diets, while even Jersey crosses have given tenderness ratings comparable with or better than those from pure beef steers. In contrast, British work has shown cross Charolais cattle to be less tender than crosses by Herefords and Devons, and it does seem that some justification does exist for Angus and Hereford claims, particularly on extensive feeding systems provided they are not exaggerated out of all proportion.

It is well established that rising pH of the meat decreases tenderness. As pH rises from 5·4 to 5·8, tenderness declines markedly but thereafter increasing pH above 6·0 leads to improved tenderness, though now associated with 'mushiness'.

JUICINESS

This is a very difficult character to assess since it relates not only to the initial impression of wetness, due to release of meat fluid, but also to the longer lasting effects of fat in the meat stimulating the salivary glands. As a result one might expect some relation between fatness and juiciness, and this is in fact the case. Contrary to popular belief fatness is a more important measure of juiciness than it is of tenderness, since it accounts for about 16 per cent of the variation in the former trait.

Moisture content, pH, age and hormones have no effect other than the degree to which age can affect fatness. Since fatness influences juiciness in a positive direction, this confuses attempts to isolate single effects. Young animals will give juicy meat, but largely because they are less fat not through age *per se*; similarly sex effects seem to be related to fatness, and bull v steer effects seem to be unimportant up to about 600 days of age.

Angus and Herefords are generally more juicy in terms of their meat than Friesians, and there is some evidence to suggest that Charolais and their crosses are less juicy than traditional British beef breeds.

FLAVOUR AND AROMA

These again are complex traits to assess, the more so since fewer people now cook and serve without spices or condiments of one kind or another, and hence disguise the original flavour of the meat to a greater or lesser degree. Since Continental housewives use more spices than their British counterparts the traits are likely to be even less important there than here. This does not mean that meat flavour is well understood. Various schools of thought exist, but it is generally felt that in the main meat flavours are derived from the fat, at least as far as species differences are concerned. Certainly fatness accounts for about 40 per cent of variation in the trait and it seems well established that flavour increases with fatness.

Age, breed, sex and hormones have only limited effect on the flavour of meat except in the case of older bulls (over 600 days), and in some studies with Charolais beef which has had less flavour than that of British breeds. These may well reflect fatness effects and in any event have not been marked. There are indications of a decline in flavour with increasing pH; still further justification for maintaining low pH values.

In general terms there seems to be some justification for the belief that the traditional beef breeds will produce better eating quality in their meat than those of dairy origin or from large Continental breeds. However this advantage is largely associated with their greater fatness, and hence is being achieved at the expense of live weight gain and feed conversion efficiency. Moreover this superior eating quality may not be readily appreciated in certain regions, particularly in EEC areas, especially if it is to be priced higher as it must be if it is to be economic to the producer. Fortunately for the Angus man the increasing affluence of Western Europe will ensure that quality meat still commands a premium, even though its overall market will not be extensive. To the large majority of the consuming public quality will continue to be defined simply as leanness, thus ensuring a continued market for cattle of large mature size and late maturity. Those British breeds which concentrate on increasing growth rate could well command a sizeable proportion of that market without losing altogether their reputation for quality, and thus benefit financially from having a foot in both camps.

Chapter 17

PRODUCTION AND MARKET PROSPECTS FOR BEEF

CHANGES SINCE THE EARLY SEVENTIES

WHEN THE first edition of this book was in preparation, immediately prior to the United Kingdom's entry into the EEC, prospects for beef were good and there was a feeling of buoyancy within the industry. There was full employment, inflation was at a very modest level compared with more recent trends, and standards of living were rising steadily. Beef was in strong demand, not only because it was the preferred meat for the majority of British people but also because retail prices were comparatively low. Certainly beef in its various forms was very much cheaper in Britain than it was across the Channel under CAP arrangements.

This fortunate situation for the British housewife was primarily due to the system of price support for key agricultural commodities that had been operating in the United Kingdom since 1947. There were no restraints on meat imports (apart from any relating to the control of disease) and the trade was able to buy meat at world prices and pass on at least a proportion of this advantage to consumers. Home producers did not suffer from this arrangement because a guaranteed price for beef was fixed at the annual Price Review. There were variations in the guaranteed price over the course of the year to take account of fluctuations in supply, for example winter and spring prices were fixed at a higher level than summer and autumn prices but over the year as a whole the guaranteed price was maintained. If farmers' returns on the open market fell below the guaranteed price for a particular week the deficiency was made good by the Government. This was not the only Government benefit for beef producers for there were various headage subsidies for breeding and rearing stock that were suitable for beef production and these included liberal payments for breeding cows on upland farms. The overall cost of the arrangements for the Exchequer, and eventually for taxpayers, was substantial but they were greatly to the benefit of consumers as well as farmers.

Prior to the war the United Kingdom had to depend to a considerable extent on imported beef and only about half of the nation's annual consumption of approximately 1·2 million tonnes was home-produced. Immediately after the war when beef and other carcass meats were still strictly rationed home-production had fallen by about a fifth from its pre-war level as a consequence of changed priorities in agricultural production, and imports, particularly from Argentina, were at a very low level.

The Government's response to this was an active policy of encouragement of domestic beef production. It was highly successful and by the early seventies total supplies were virtually back to the pre-war tally of 1·2 million tonnes but with this difference, approximately 85 per cent were attributable to home production. In the meantime the human population had increased appreciably but beef consumption at about 24 kg per head was within 5 kg of the pre-war figure. It was still the most popular meat with the various forms of pig meat in second place at less than 20 kg per capita. Comparatively, however, the average Briton was only a modest consumer of beef for the average in the United States at that time was just over 50 kg per head while Argentina, the leader of the beef-eating league had an average consumption at least twice that of the Americans.

Buoyancy in the industry was not maintained after entry into the Common Market and housewives have suffered as well as producers because target prices for beef had been set at levels appreciably above world market prices. The consequence is that consumer resistance has developed to weaken the demand for beef, a point which will be discussed in greater detail later. The higher Community prices for beef failed to bring any long term benefits to producers for their costs have risen appreciably and there was also the loss of some of the direct subsidies, in particular that for calf rearing. The impact of the changed situation has been most marked in the suckler calf sector of the industry. In 1974 there were over 1·9 million beef breeding cows in the United Kingdom, which was the highest ever tally but by 1983 there were half a million less, a drop of approximately 30 per cent.

Measures were eventually taken to arrest this decline. A breeding cow premium of £12·37 per head was awarded in 1981 provided there were no milk sales from a farm. In the same year the Hill Livestock Compensatory Allowance of £44·50 was introduced. All breeding cows, both dairy and beef, qualify for this award provided farms are located in what are rather blandly described as 'less favoured areas'. These allowances may appear to be generous, but taking into account how inflation has eroded the value of the pound since 1970 they do not compare with the subsidies operating prior to joining the EEC. There is no evidence to suggest that they have yet halted this decline in beef breeding cow numbers from their 1976 peak.

Another more recent EEC policy change could also affect the supply of cattle for beef production and this is the compulsory reduction of milk sales from United Kingdom dairy farms, imposed in 1984 and amounting to approximately 8 per cent of the 1983 level of production. At the time of writing it is not clear how the majority of dairy farmers will react to this limitation but immediately there has been considerable publicity about reducing herd numbers and increased slaughtering of dairy cows. One must hope for the future of beef production that this will not be the general reaction of dairy farmers and it need not be the case. British dairy farmers tend to be liberal users of concentrates and this along with their general preference for Friesians is a principal reason for the high average yields that are now recorded. But the law of diminishing returns applies in the feeding of dairy cows and the last increments in milk production can be very expensive on this account. Dairy farm costings, such as those undertaken by the English Milk Marketing Board, reveal that there are factors other than high yields that determine profit. Many farmers could be better off with lower milk production per cow and lower cake and fertiliser bills without any need to reduce herd size.

However where there are herd reductions there is little reason for believing that any slack in a farm's carrying capacity will be taken up by suckler cows as the Minister of Agriculture rather fatuously suggested when the quotas were first announced. For many farmers a better proposition would be the retention of a proportion of the annual crop of bull calves for rearing rather than selling them all at the week-old stage. Ironically the imposition of milk quotas is going to act as a deterrent against farmers going out of dairying and into beef as some have done in recent years, encouraged by a so-called golden handshake for leaving the industry. It is most unlikely that many farmers will wittingly sacrifice the considerable asset represented by their milk quotas by giving up dairying.

SUPPLY AND DEMAND PROSPECTS

Taking the EEC as an entity total beef supplies have been remarkably stable over recent years though there has been some decline from the 1980 peak figure of 7·17 million tonnes. Apart from trading between member states there are imports of the order of 350,000–400,000 tons annually but these are more than balanced by beef exports to non-member countries. Much of this imported beef comes out of intervention holdings. This is the EEC device for controlling prices which involves putting surpluses over current demands into store and their release when beef is in short supply. That is the theory but beef that comes out of storage is a somewhat different article from the beef that went into store in the first place and much of it is destined for manufacturing or for export, principally to North African and Eastern European countries including Russia.

Costs of storage and deterioration in store make the whole operation expensive. In 1981 EEC expenditure on direct support measures for beef was approximately £800 million and national governments made further contributions. It is little wonder that the demand for beef is declining with this situation for the costly intervention and export subsidy schemes that operate are designed to maintain high internal prices by preventing the interactions of changing supply and demand situations. Under the deficiency payment system that the United Kingdom operated before joining the EEC housewives reacted to increased supplies at lower prices by buying more beef. Not only did they enjoy the benefit of cheaper meat but there were no storage costs or losses due to deterioration in store.

Despite keeping beef prices at an artificially high level and the massive contributions coming from the EEC and member governments in respect of their own farmers, beef production is not and probably never can be a profitable enterprise in countries where land is expensive and holdings are comparatively small. Specialised beef production if it is to be profitable has to be organised as a low input enteprise on broad acres. Indeed it is surprising that so many United Kingdom farmers have continued to produce beef and that the decline in its production since the mid-seventies has not been on an even greater scale.

Unless there is an economic revival with a return to much fuller employment it is unlikely that there can be any improvement in beef prices apart from any that is attributable to inflation. Looking at the situation from the consumers' viewpoint there is still a strong preference for beef despite the fall in per capita consumption that has recently put beef into second place after pigmeats. Consumer studies, reported by the MLC, reveal that housewives consider it to be too expensive and considerably poorer value than either poultry or pork. There is little doubt that the better cuts of prime beef are now luxury items for the typical housewife working on a tight budget. If beef is more than a very occasional choice for her it will probably have to take the form of stewing steak or mince-meat. It is a sad thought for those who remember traditional roasts of beef with their accompanying Yorkshire puddings. Perhaps this deprivation is part of the price that British people will continue to pay for the privilege of membership of the Community until such time as there is a drastic and long overdue revision of the Common Agricultural Policy.

INDEX